Packaged Air Conditioning

Packaged Air Conditioning

B.J. Hough, FCIBSE, FInstR, MCIM

Butterworth-Heinemann Ltd
Linacre House, Jordan Hill, Oxford OX2 8DP

  PART OF REED INTERNATIONAL BOOKS

OXFORD   LONDON   BOSTON
MUNICH   NEW DELHI   SINGAPORE   SYDNEY
TOKYO   TORONTO   WELLINGTON

First published 1993

British Library Cataloguing in Publication Data

A catalogue record for this book is
available from the British Library

ISBN 0 7506 0920 6

Library of Congress Cataloguing in Publication Data

A catalogue record for this book is
available from the Library of Congress

Typeset by Vision Typesetting, Manchester
Printed and bound in Great Britain by
Biddles Ltd, Guildford and King's Lynn

Contents

Contents

Preface

The advantages of controlling the internal environment are well known and have been practised for a number of years, but are thought by some to be complicated and to an extent unnecessary.

Indeed, it has only been in the past 30 years that control during the winter by central heating has been a necessary feature of UK homes. A similar revolution is now taking place with regard to summer environmental control or 'comfort cooling' in commercial premises. Such a term is imprecise, in that the systems dehumidify as well as cool. In some instances, for example a shop, the number of people contributing to the cooling load and the need for effective lighting could result in the peak load being at its highest in the winter, around Christmas, rather than the summer.

When buildings had high ceilings, small windows and high ventilation rates due to excessive leakage and low occupation densities, then overheating problems were minimal. However, today's buildings are not constructed or occupied as were those in Georgian times. To conserve energy, the insulation of the building fabric has been increased. Occupation densities have nearly trebled because of the high cost of commercial space, and modern business practices including information technology have increased the internal gains. All this adds up to a serious overheating problem in many business premises.

With such internal conditions one could almost forget those

outside which nevertheless contribute to the problems inside. Solar radiation can be a problem because of lower sun altitude compared with that in the tropics. The rays which track through vertical windows add as much as $400 \, W/m^2$ to those on the south facade. Even traffic in city centres will create microclimates where a car will give off enough heat to heat about three average-sized homes. Although air conditioning may not be the complete answer to these problems, it can help to alleviate the worst situations and make the atmosphere in our buildings more tenable. Such air conditioning does not need to be 'full' air conditioning, where the internal temperatures and humidities are controlled both summer and winter together with providing outside air for ventilation and various forms of filtration to treat outside and inside pollution.

A plant can be designed only to deal with overheating and high humidity where these are the worst of the problems and may supplement conventional heating and ventilation solutions. It is these areas that this book seeks to address, while taking into account the availability of modern packaged or pre-engineered equipment.

Although this equipment makes the application of air conditioning easier and more economical, to use it to full advantage a proper analysis needs to be made to reach a vital understanding of the way in which these units function.

The equipment, which is produced by high-volume production techniques, gives a cost-effective means of providing a better thermal environment in modern buildings. Frequently, by decentralizing the plant, any distribution losses which occur with central plants are also minimized.

Even though the equipment is pre-designed and pre-engineered, the range of sizes available can accommodate most applications with sufficient flexibility and cope with all but those applications which require very close control. However, to make the best and most efficient use of equipment and to match it to the many modern applications requiring environmental control makes a knowledge of equipment design essential. The better the match, the more economic will be the initial cost and subsequent running costs.

The book is aimed at those who wish to learn how to apply simple systems to single-zone applications, particularly those in allied fields such as building and mechanical or electrical contracting. Some of the information can be used for larger installations and there is therefore a chapter which gives details of packaged air conditioning systems which are suitable for larger multizone application. It is a practical work outlining in clear, simple terms subjects of apparent complexity. It can also be useful to those who have to specify air conditioning for small applications, so that they can better understand and if necessary check the systems offered. Since it is not always possible to locate individual units within the space to be conditioned, a ducted distribution system may be necessary; therefore a separate chapter is included giving approximate methods of duct sizing.

Air conditioning is a very complex subject with many specialized areas requiring specific knowledge and experience. This book, in dealing with simplified packaged air conditioning, is intended to be a stepping stone to an understanding of the full design process and analysis which may be necessary where the application is large or complex or requires special conditions. Many of the subjects covered require high levels of academic ability, up to degree level, and years of experience to put them to proper use in practice.

Acknowledgement

My thanks are due to John Allen of Airconair Services Ltd, a great engineer and a greater friend, for his help, guidance and constructive criticism during the writing of this book.

ONE

Comfort

Assessing comfort

The purpose of air conditioning is to control the internal environment, either for the needs of the occupants of a space or to meet the requirements of the heat-producing equipment of the process carried out within that space. Although it is the occupiers of the space and their comfort that matter, people have different ideas as to what constitutes comfort. Individuals react differently to the same levels of temperature and humidity and find it difficult to express their comfort needs. They therefore find comfort difficult to define, except as a feeling of well-being. Most people will express their needs in terms of discomfort, by saying they are too hot, too cold, too muggy, and so on.

A further problem in assessing comfort is that activity and the amount of clothing worn affects a person's reaction to the environment. For example, greater activity increases the amount of heat to be lost and the air temperature and humidity should be reduced to maintain comfort. Clothing insulates the body and keeps in the heat, and since men and women wear different amounts and types of clothing they will react differently to the same conditions. Owing to their different metabolism and to different clothing factors, men will generally be happier with a lower temperature than will women.

The air conditioning of a space is correct when you cannot see it, feel it or smell it; in fact you should have no sensation other than that of well-being. It is rare for this ideal to be achieved, but to create conditions where the occupants can devote their attention to their appropriate activity and not use energy in fighting the environment should be the designer's aim.

To achieve the right conditions with respect to the thermal environment you need to know how the body gains or loses heat, what physical factors control the loss or gain and how an air conditioning system achieves this control. To be comfortable and to function effectively, the body needs to lose heat at a balanced rate to suit outside influences. If it loses it too fast, the body experiences cold; if too slowly, the body will feel overheated.

How body heat is transferred

The body gains or loses heat in four different ways, by conduction, radiation, convection and evaporation.

Loss by conduction is by direct contact with a cold surface, such as the effect of placing one's feet on a cold floor.

Loss by radiation is when the body radiates heat to a colder surface, such as a cold window. Conversely, heat is gained by radiation from a hot source, such as the sun or a radiant fire or high lighting levels.

Loss by convection is by colder air passing over the body and absorbing heat. Heat can be gained by convection, but for this to happen the air temperature needs to be higher than the body surface temperature.

Finally, loss by evaporation is through breathing and by perspiration from the pores. If beads of moisture appear on the skin, then the body is unable to lose sufficient heat by evaporation and discomfort occurs.

Controlling heat losses

Conduction loss is reduced by insulation either by a carpet, for example, or by thicker soles on shoes. Radiation loss is minimized by placing a shield between the source of radiation and the body, such as an outside awning which will shield the sun's rays or a curtain reducing the cold radiation effect of a window in winter.

Both of these losses are best cured by a primary physical means rather than by a secondary means such as an air conditioning system. However, comfort can still be provided by negating the effects by balancing the amount of heat which is lost or gained by convection and evaporation. A high radiant gain can be counteracted, to an extent, by increasing the convective or evaporative loss. This is the job of the air conditioning system – by controlling the air temperature, air movement and relative humidity within a space.

The rate of convection is affected by air temperature and air movement. The higher the temperature, the less the convective loss and the higher the air movement, the more the convective loss, and vice versa. Increasing the convective loss therefore compensates for the radiant gain, provided that the air movement is draught free.

Evaporation is affected by air movement and relative humidity, a measure of the effect of moisture in the air. The higher the relative humidity, the less heat is lost by evaporation; the higher the air movement, the more heat is lost by evaporation. Again, an increase in the evaporative loss will compensate for a radiant gain.

Achieving comfort

Although people do have differing comfort levels, it is found that most will be reasonably comfortable in a summer temperate climate with a room temperature of 23°C and a relative humidity of 50%. As their perception of change in the thermal environment is not good, the temperature can range between 21°C and 25°C and the relative humidity between 35% and 65%. Discomfort will only be sensed at the extremes of this range and when comfort factors are unbalanced.

The air conditioning system achieves comfort by changing the condition of the air supplied to maintain room conditions, and controlling those conditions by removing unwanted heat and moisture occurring within the space.

Conditions other than the temperature and humidity, such as air quality, can also affect comfort. An air conditioning system can help to maintain the quality of the air by providing the right amount of filtered outside air to replace the stale air in any given space. This will reduce the contaminants in the air, such as tobacco smoke and odours. The type of filter and the amount of replacement air will depend on the activity and degree of contamination. Normal filters, made from expanded poly-urethane or woven fibres, will reduce dust content, whereas electrostatic filters will reduce smoke and pollens and the use of activated charcoal will reduce some smells. Alternatively, these requirements of ventilation and air cleanliness can be achieved by separate systems working in conjunction with the packaged air conditioning system.

Comfort cannot be properly theoretically defined, as it varies from person to person depending on the time of day and other personal circumstances. As perception of temperature and humidity is not exact, trying to achieve exact conditions is unnecessary as well as being costly. The temperature and humidity can be allowed to float within a range and this should be made clear to the occupiers of the space to be conditioned. Always ensure that expectations of air conditioning are not so high that ideal conditions are anticipated for each person.

Building-related illness

One of the areas of concern related to air conditioning but not strictly to comfort is building-related illness. The foremost of these problems are Legionnaire's disease, humidifier fever and sick building syndrome. These illnesses, which have been largely associated with central plant, are unlikely to occur in most packaged air conditioning systems which are designed to over-come a specific problem and to a large extent are decentralized. Nevertheless the user still has a duty to ensure proper mainten-ance which will further reduce or eliminate any risk.

Legionnaire's disease

Legionnaire's disease is particularly related to water sprays when water within a particular temperature range becomes contaminated with the micro-organism *Legionella pneumophila*. The organism occurs naturally and is not harmful under normal circumstances. It needs to multiply in contaminated water to cause problems. At temperatures below 20°C it will hardly multiply, and above 60°C it dies; therefore the ideal water temperature for the organism is around 40°C which does not normally exist in the water associated with packaged air conditioning. The concentration must then be sprayed in the form of an aerosol which can be breathed in by a susceptible person for the disease to happen.

The majority of cases of Legionnaire's disease on record have been due to sprays from showers or hot water installations. Those related to air conditioning have been because of contamination of cooling towers in central plant systems. Most packaged systems are air cooled so that the situation does not occur. The only water usually present within the system is that of the condensate which is not usually within the temperature range mentioned above and is of such small quantity as to make it virtually impossible to create an adequate aerosol spray to pass on the micro-organism.

Humidifier fever

Humidifier fever is not an infection but an allergic reaction to certain micro-organisms which can grow in the stagnant water of the humidifier's reservoir when the unit is not in use, usually at weekends. When the humidifier starts up after a weekend, the concentration of micro-organisms is sprayed into the space causing a reaction of headaches and flu-like symptoms. As the concentration is diluted by replacement water while in use, the reaction diminishes until it is hardly apparent by the end of the working week. Unlike Legionnaire's disease, if those affected are removed from the contaminated space the reaction stops.

Most packaged units are designed only to dehumidify rather

than humidify which is a function usually needed in winter. Humidifiers are fitted to close-control packaged units and are normally steam humidifiers and with continuously flushing cycles so that the problems of humidifier fever are unlikely to occur. The problem only seems to occur with evaporative spray humidifiers having a water reservoir.

Sick building syndrome

Sick building syndrome is the most difficult to understand as there seems to be no one cause that can be identified as the main reason for its incidence. Insufficient ventilation, poor air quality, inadequate maintenance and lack of individual control have all been cited as contributory factors. To date, most situations have been related to central plant in large buildings. Part of the problem could be caused by people expecting air conditioning to provide the ideal environment for them and when it does not, dissatisfaction occurs. Whatever the reasons, the syndrome does not seem to happen in the smaller installations using packaged units. Maybe the expectations of the occupiers are lower and their apparent control over their system greater.

TWO

Load estimation

Correct load assessment

The load affecting air temperature is from the sun, transmission through the building structure and internal sources such as equipment, lighting and people, whereas the load affecting humidity is mainly from ventilation air and people.

The key to any good air conditioning system lies in the correct analysis of a building's load. The size of the plant, distribution and terminal devices, together with the control of the complete system, depends on this analysis. Estimating a load which is too small will lead to dissatisfaction by the occupants, whereas overestimating will lead to high capital costs, rapid cycling and therefore inefficient plant operation, and unnecessary maximum demands for power. By knowing all the factors which have been taken into account during a proper analysis, the designer will be able to make correct judgements in selecting packaged equipment.

Without a thorough experience of proper analysis, 'rules of thumb' should be avoided. Estimation procedures need not however be unnecessarily complex or time consuming, but should consider the size of the building and the need for accuracy.

There are ranges of computer programs available which can analyse the building's requirements for the many changing conditions and states of a building for different times and ambient conditions. Although these are ideal for large complex central plant systems in multistorey office blocks, using them on single-zone packaged applications is like using the proverbial sledgehammer to crack a nut.

As any building is unlikely to be built exactly as planned, and certainly is rarely used as the designer envisaged, the high degree of accuracy and the complexity of a dynamic program is rarely necessary, especially when considering packaged air conditioning systems. Less, too, is the need to consider extensive diversity over the whole building.

Factors affecting load

Think of a building as a big storage vessel absorbing heat from areas of high temperature and giving out this heat as the temperature decreases. It absorbs heat from the sun, the surrounding air and spaces known as transmission heat, and the energy released inside the space from lights, people and equipment. Included in the heat by transmission will be the load caused by the infiltration of air into the space or the need to provide additional ventilation air.

Solar radiation

The heat from the sun affects the building in a number of ways. It can be transmitted by direct radiation from the sun to an object directly in its path, or by indirect or diffuse radiation, where some of the direct rays are deflected or diffused by small droplets of moisture or motes of dust in the air. Thus the effect of solar radiation is felt on a cloudy day or by the absorption of heat from the sun, even on a north-facing facade.

In the Tropics, where the sun can be directly overhead, its rays on horizontal surfaces are more intense than in more northerly latitudes. In this situation a well-insulated roof with large overhangs or eaves capable of reflecting the rays provides part of an effective solution to the problem of overheating. In northern latitudes such as the UK, the sun's altitude in the sky is lower, affecting vertical surfaces such as walls and windows which prove more difficult to shield. An overhanging roof is less effective and exterior blinds or awnings need to be considered to provide the answer. In the UK the problem can be as acute in March and October, when the sun's altitude is even lower, as it is in June. Roof glazing in any climate always merits special consideration.

Untreated glass is virtually transparent to solar radiation and barely absorbs any of the sun's energy. The sun's energy passes through the glass, heating up any solid object in its path, such as the floor or desks, or it is reflected to heat up other surfaces within the space. This absorbed heat is then emitted to the air after a period dependent upon the mass of the object. Being solid, exterior walls of the building and the roof absorb the heat, transferring it through the structure to the inside and causing a similar delay to that which has been absorbed from the rays through the glass. This has a tendency to smooth out the peaks and troughs of loads.

The effect of the sun's rays through glass can be reduced by shading. This shading can be by means of exterior blinds or awnings or purpose designed into the building structure. It is unlikely that there will be a significant reduction from internal blinds because once the sun's rays have passed through the glass it is very difficult to reflect the energy back, even when the blinds are of reflecting material. They do, however, reduce direct radiation on the occupants and the discomfort from solar glare.

The walls and roof of a building may also be shaded, but this is usually due to the shade from an adjacent building. Consider only external shading and the time at which the shading occurs. This latter factor is important, as shading which occurs in the morning will have little effect if the maximum load occurs in the afternoon. Another consideration for reducing the effect of solar radiation through glass is by the use of heat-reflecting or heat-absorbing glass or the use of special solar film.

The building's compass orientation or the azimuth angle of the sun could both affect the load assessment. Surfaces facing east will give a load in the morning, whereas those facing west will give a load in the afternoon. Thus some applications may not require peak design cooling throughout the day.

The air conditioning system deals with this solar load when it has been absorbed and then released by convection into the circulating air.

Part of the transmission loads on the building will be those of conducted heat due to the effect of the higher temperature of the outside air. They are caused by the microclimates which occur in our town centres and are often 10°C higher than that recorded at weather stations and will affect both the sensible and the latent heat requirement. Another part of the transmission load could be from adjoining property through the party wall or from unconditioned areas through the internal partitions. If the load from such transmission is due to the activity in an adjacent space, then this must be included in the assessment.

Ventilation requirements

The ventilation load also has a significant effect on the cooling load estimate. The amount of ventilation required depends on the needs of the occupants and controlling the level of contamination in the room. For comfort air conditioning, it is more generally related to the number of people occupying the space and the level of smoking, as this is considered the major contamination from people. However, with today's attitude to smoking careful consideration of the application needs to take place to avoid overestimating the degree of pollution. For example, at one time a restaurant was considered an area of heavy smoking, but now most restaurants restrict the area of smoking and this area may need separate treatment. Otherwise the amount of ventilation air should be that for 'some smoking' rather than 'heavy smoking'. There could also be a ban on smoking, say in an office, when the ventilation requirement could be that for 'no smoking'.

Sometimes the amount of air which infiltrates into a room naturally through cracks or openings is sufficient for all the ventilation requirements such as private offices or small shops. This is unlikely in restaurants, conference rooms, general offices and larger shops, and therefore additional ventilation air will need to be provided.

This infiltration air will vary according to the structure. Modern office buildings which are double glazed and well insulated are unlikely to have an infiltration rate of more than one air change, whereas an older building with single-glazed sash windows could have a rate of over two air changes per hour. In simplifying the load estimate table it was not possible to take a wide range of natural ventilation into account, so the chart used tends to underestimate the ventilation load for private offices and small shops. It may be necessary to use more specific data if the air change rate or extraction is known.

Internal loads and conditions

The internal loads are those from people, lighting and heat-producing equipment such as office machines or catering equipment. These loads can often be so high that they can create a need for air conditioning even when the outside temperature is as low as 5°C.

The mass of the building will have an effect on the actual peak cooling load, which can occur in unoccupied periods when the plant would normally be switched off. This stored heat could be retained overnight by the building mass and will therefore need to be removed when the plant restarts. As the outside air temperature is likely to be lower than the inside temperature overnight, there may be scope for what is called 'free air cooling'. This could also be used when cooling is required in winter.

The problem is one of effective control. In summer the temperature difference may not be sufficient to cool the room and certainly it will have little or no effect on the high humidity likely to be present. In winter there is the need to ensure that any cooling by outside air is controlled without using supplementary heating, or this may be greater than the energy saved by cycling off the compressor. Again there will be no control over the humidity; with ambient winter air and the usual internal humidity conditions, the result could be unacceptable levels.

Not all the instantaneous load on a wall or roof will eventually appear as a load within the space. The radiant heat from the sun's rays will raise the surface temperature above that of the surrounding air, so that some of the heat is returned to outside air.

The sun moves across the sky, affecting different parts of the building with varying intensity, and at the same time the temperatures and humidity are also changing causing the cooling load to change. This dynamic situation is difficult to calculate even when the internal loads do not vary. There are computer programs which can assess these dynamic conditions, but for most small system requirements a simple approach is sufficient, by calculating the requirement assuming that steady-state conditions apply, or where solar radiation is considered at a specific time or series of times and all other loads are considered as constant. Again there are computer programs available that can make these calculations, ranging from one that considers the sun's position for every hour throughout the year, to a very simple program based on a pocket calculator dealing with a specific time and set of conditions.

For the purposes of packaged units an assessment based on steady-state conditions prevailing in the building is sufficient. This can be an assessment against a single average set of conditions, or those occurring at say five representative times and with varying outside temperatures depending on location. What it should consider is the different loads which can occur in a building. These loads will either be sensible heat or latent heat. Sensible heat affects the temperature of a fluid, whereas latent heat affects the state that is the change from liquid to vapour. Thus sensible heat loads increase the air temperature and latent heat loads increase the humidity.

Often there could be a situation when loads occur at different times, such as in a restaurant where the maximum external load from the sun occurs during the day but the maximum internal load is in the evening when the external load is much lower. These matters should also form part of your assessment.

Method of calculating cooling load

Figure 2.1, Table 2.1 and the method below are based on steady-state conditions in a single-zone application of no more than 300 m² and on an outside temperature of 27°C db (dry bulb) and 20.5°C wb (wet bulb) with an inside temperature of 23°C db and 50% RH, but adjusted so that allowance is made for both solar and transmission gains for each aspect considered. The figures are average for season and location, but will give an assessment which will be suitable for most buildings in the UK when considering the application of standard small package air conditioning for comfort. Unlike many rules of thumb, the method can provide information on both the sensible heat and latent heat requirements, enabling you to select a more suitable unit. The selection should not be based on the total capacity only, but on information provided by the manufacturer about the sensible and latent capacities of the equipment.

To complete Figure 2.1 proceed as follows:

1 Enter the name and address of the building being considered.
2 Enter the building type, i.e. the application and the net length, width and height of the zone or space to be conditioned.
3 If more than one face of the zone is sunlit during the day, pick only the one with the worst load. All others are to be considered as if they were in the shade.
4 Calculate and enter the area of the glass on the sunlit face.
5 Calculate and enter the area of glass on all other exterior sunlit faces as in the shade.
6 Calculate the gross area of the wall on the sunlit face, subtract the area of glass in 4 above, and enter the net area.
7 Calculate the gross area of the other exterior walls, subtract the area of glass in 5 above, and enter the net area of the wall.
8 Calculate the gross area of internal walls and enter as internal walls.

Packaged Air Conditioning

CUSTOMER		BUILDING	
NAME		TYPE	
ADDRESS		LENGTH	
		WIDTH	
		HEIGHT	

SURFACE	AREA or No.	FACTOR	SENSIBLE HEAT	LATENT HEAT
GLASS: Sunlit Shade				
WALLS: Sunlit Shade Internal				
ROOF or CEILING				
FLOOR				
LIGHTING				
APPLIANCES				
PEOPLE: Sensible Heat Latent Heat				

VENTILATION – No. People	Requirement	Factor		
			TOTAL	

Figure 2.1 *Load estimation chart, for comfort conditioning*

14

Table 2.1   Load estimation factors for comfort conditioning*

If there is more than one aspect, take only the main sunlit face; all others are considered in the shade		

Glass: plain single glazing (W/m²)	*Bare*	*Outside awnings*
N or in the shade	95	
NE or NW	158	95
E or W	380	237
SE or SW	440	283
S	380	273

Note: For double glazing reduce by 10% and for special solar ban glass reduce by 40%

Outside walls (W/m²)	*Uninsulated*	*Insulated*
N or in the shade	11	6
NE or NW	14	10
E or W	25	19
SE or SW	22	13
S	30	22

Flat roofs or sloping plan area only (W/m²)

	Uninsulated	*Insulated*
	15	10

All internal walls, floors or intermediate ceilings (W/m²)

8

People (W/person)	*Sensible heat*	*Latent heat*
Office	91	50
Shop	100	60
Restaurant	110	70
Light work in factory	114	120
Medium work in factory	125	140

Ventilation air (W/m³/s)	*Sensible heat*	*Latent heat*
	4600	9600

Ventilation requirements (m³/s)	*No smoking*	*Some smoking*	*Heavy smoking*
	0.008	0.012	0.024

*This simplified design procedure is for single-zone applications of 300 m² or less.

9 If there is a roof above the zone, enter the area of this roof in 'Roof or ceiling'. However, if this is an internal zone or space, then it is considered a ceiling. Flat roofs with a false ceiling below are considered as roofs. If part is roof and part is ceiling enter each area separately. If there is a skylight or other roof glazing, then take this area separately and use the factor for south-facing glass.

10 Calculate and enter floor area.

11 Enter the load relating to lighting. This information may be given in terms of watts per metre of floor area, in which case enter floor area. It may alternatively be given as the number and rated wattage of the lights. Remember that with fluorescent lighting the heat given out is about 25% more than the tube rating. Therefore, if the tube indicates 100 W then the heat given off will be 125 W.

12 Enter the load from equipment or appliances. Remember that many appliances will operate under normal conditions at less than their rated wattage. Also, the equipment may not be operating all the time. Therefore, allow for diversity when making this assessment.

13 Enter the number of people within the space. If the number is the maximum, then again allow for possible diversity.

14 Again enter the number of people to allow for the ventilation load to be considered.

15 From Table 2.1 select the appropriate factor for glass and sunlit orientation and multiply this by the area, to give the sensible heat through the unshaded glass.

16 From Table 2.1 select the factor for shaded glass and multiply this by the area, to give the sensible heat through shaded glass.

17 From Table 2.1 select the appropriate factor for walls and sunlit orientation and multiply this by the area, to give the sensible heat from sunlit walls.

18 From Table 2.1 select the factor for walls in the shade and multiply this by the area, to give the sensible heat from exterior shade walls.

19 From Table 2.1 select the factor for internal walls and multiply

this by the area, to give the sensible heat through internal walls and partitions.

20 From Table 2.1 select the appropriate factor for either roof or ceiling or both and multiply by the required area or areas, to give the sensible heat from the roof or ceiling or both.

21 From Table 2.1 select the factor for floors and multiply this by the area, to give the sensible heat from the floor.

22 In the 'Factor' column put down the W/m given and multiply by the area, if appropriate, to give the sensible heat from lighting. Alternatively, enter 1 for incandescent lighting or 1.25 for fluorescent lighting and multiply by the total wattage of all the lights, to give the sensible heat.

23 Enter the rated wattage of the equipment, adjusting for diversity if considered necessary, giving the sensible and maybe the latent heat from the equipment or appliances.

24 From Table 2.1 select the factors for sensible and latent heat from people relating to the application and multiply each by the number of people, to give both the sensible and latent heat from people. Adjust for diversity if considered advisable.

25 From Table 2.1 select a ventilation rate applicable to the amount of smoking estimated in the conditioned space and multiply by the number of people and the factors for sensible heat and latent heat.

26 Total the 'Sensible heat' column.

27 Total the 'Latent heat' column.

28 Add the totalled sensible heat to the latent heat to determine the total heat for the space.

For most applications there is a relationship between sensible and latent heat, and such a relationship can be used as one of the checks to establish if the load estimate appears right for the application. Particular applications will have specific proportions of sensible to latent heat which can be shown as a sensible heat ratio (SHR). This is derived by:

$$\text{SHR for the room} = \frac{\text{Sensible heat from load estimate}}{\text{Total heat from load estimate}}$$

For example, a private office would normally have only one person occupying above-average space and therefore the sensible heat ratio will be high, say about 0.95. If more people occupy the same area, the ratio will fall. Conversely, a restaurant would have many people occupying the same floor space, with additional moisture coming from the food. This would result in a lower sensible heat ratio of, say, 0.7.

Table 2.2 gives a range of room sensible heat ratios for different applications. These values can be used to check the load calculations to establish if they are correct or to determine the proportions where the calculations give only the total heat load for a given application.

Table 2.2   Room sensible heat ratios*

Application	Ratio range
Private offices	0.95–0.90
General offices	0.85–0.80
Small shops	0.85–0.80
Large shops or departmental stores	0.80–0.75
Restaurants	0.75–0.70

*Check your load calculations if the range is below 0.65.

THREE

Psychrometrics

Thermodynamics of moist air

Psychrometrics is about the thermodynamics of moist air. When the temperature of air changes, other characteristics of the air change at the same time. A knowledge of these changes is necessary to ensure that the right conditions are obtained in a space. If there is a need to remove heat and moisture from a room, then air which is cooler and drier than that of the room needs to be introduced. How much cooler and drier depends on the amount of sensible and latent heat which must be removed. Sensible heat will affect the air temperature in the room and latent heat the air's moisture content. The right quantities of each needs to be removed and this is determined by psychrometry.

However, before considering the ways in which the air can be changed, a knowledge of its various properties is necessary. The properties of moist air are as follows:

Dry-bulb temperature (°C db)
Wet-bulb temperature (°C wb)
Relative humidity or percentage saturation (%)
Moisture content (kg/kg)
Specific enthalpy or total heat (kJ/kg)
Specific volume (m³/kg)
Dew-point (°C)
Pressure (Pa or bar)

The psychrometric chart

The properties listed above can be depicted graphically on a psychrometric chart. The last property is that of barometric pressure, and is measured in Pa or bar. Each psychrometric chart is for air at a particular barometric pressure or for specific height above sea level. For standard air charts in the UK this is normally taken as at sea level.

Dry-bulb temperature
Dry-bulb temperature (°C db) is the temperature of moist air measured with a thermometer having a dry sensing element. Read horizontally from left to right on the chart in Figure 3.1.

Wet-bulb temperature
Wet-bulb temperature (°C wb) is the temperature of moist air measured with a thermometer having a permanently wetted sensor by means of a wick and which is rotated in the air. Read at approximately 35° diagonal from right to left on the chart in Figure 3.2.

Relative humidity
Relative humidity (% RH) is the ratio of the actual vapour pressure of the moist air to the vapour pressure were the air saturated at the same temperature. Read from the concave curved lines from 100% to 10% on the chart in Figure 3.3. All relative humidities are temperature dependent.

Percentage saturation (%) is the ratio of the moisture content of moist air at a given temperature to the moisture content of saturated air at the same temperature. It is very similar to relative humidity and is often used in place of it. Read from the concave curved lines from 100% to 10% on the chart.

Moisture content
Moisture content (kg/kg) is the actual weight of water vapour present per kg of dry air. Read vertically from the bottom up on the chart in Figure 3.4. This is the best indicator of the change in actual moisture content of the air by the air conditioning process.

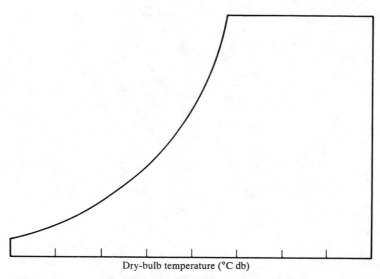

Dry-bulb temperature (°C db)

Figure 3.1 *Dry bulb temperature on psychrometric chart*

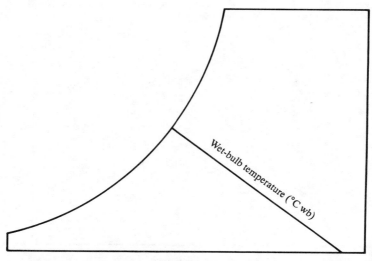

Wet-bulb temperature (°C wb)

Figure 3.2 *Wet bulb temperature on psychrometric chart*

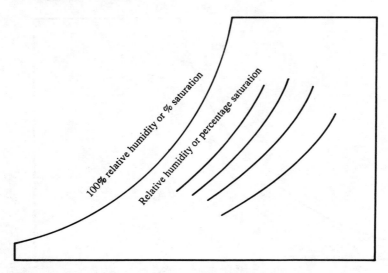

Figure 3.3 *Relative humidity on psychrometric chart*

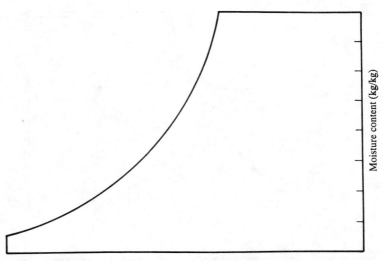

Figure 3.4 *Moisture content on psychrometric chart*

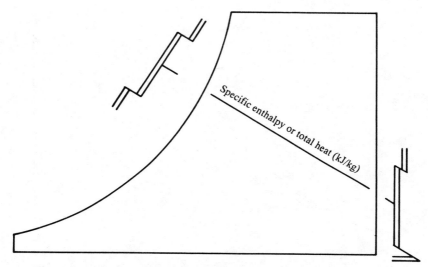

Figure 3.5   *Specific enthalpy on psychrometric chart*

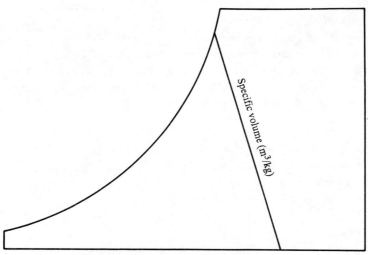

Figure 3.6   *Specific volume on psychrometric chart*

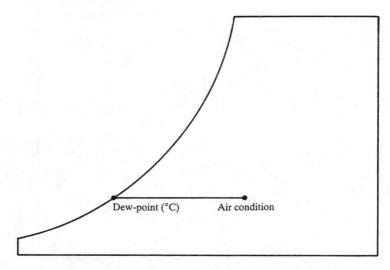

Figure 3.7 *Dew-point temperature on psychrometric chart*

Specific enthalpy
Specific enthalpy (kJ/kg) is the total heat per kg of dry air above a datum of 0°C. Read between scales off the right and bottom of the chart in Figure 3.5 across to scales to the left and top of chart. This is the value used for the determination of the actual total heat capacity of equipment.

Specific volume
Specific volume (m³/kg) is the volume occupied by 1 kg of atmospheric air. Read at approximately 60° diagonal across the chart in Figure 3.6. This enables the air volume required to carry a given amount of heat to be determined.

Dew-point
Dew-point (°C) is the temperature at which the vapour pressure of a sample of moist air is equal to the saturated vapour pressure.

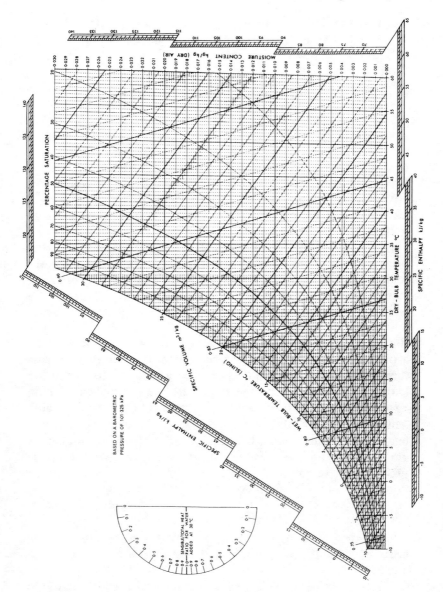

Figure 3.8 *Complete psychrometric chart for UK – based on a barometric pressure of 101.325 kPa (By courtesy of the Chartered Institution of Building Services Engineers)*

Read at the 100% percentage saturation line on the chart in Figure 3.7. This is the temperature at which any water in the air will appear as dew or vapour.

Uses of the psychrometric chart

For most purposes the complete psychrometric chart (Figure 3.8) will give sufficient accuracy for design calculations. However, if greater accuracy is required, then these calculations may be obtained from tables given in standard reference books such as the Chartered Institution of Building Services Engineers' Guide. The advantage of the chart is that various processes and conditions of air can be depicted graphically and are therefore easier to understand.

Knowing two of the properties of air at a particular barometric pressure will enable a point, representing the condition of the air, to be plotted on the chart. From this, all other properties of the air can be determined. For practical purposes the two properties normally taken to establish the condition of the air are the dry- and wet-bulb temperatures. These are measured using either a sling psychrometer, also known as a whirling hygrometer or an aspirated hygrometer. The first consists of two mercury-in-glass thermometers mounted so they can swing around a handle rather like a football rattle. One thermometer, as a small wick or sock, covers the bulb which can be wetted to give the wet-bulb temperature. The other bulb is uncovered to give the dry-bulb temperature. The aspirated hygrometer has the two thermometers mounted in a tube with a clockwork-driven fan drawing air up the tube and across the thermometers.

Psychrometric processes

The majority of processes associated with air conditioning are heating, humidification, cooling, dehumidification and air mixing.

Heating can be depicted by a horizontal line drawn from left to right of the chart in Figure 3.9. The dry-bulb temperature increases along the bottom scale and the moisture content remains unchanged on the right-hand vertical scale.

Humidification, which is adding moisture to the air, can be shown as a vertical line from bottom to top provided that no dry-bulb temperature change takes place. Therefore, moisture content increases on the vertical scale with no change on the bottom scale. If, however, as in the case of a spray humidifier, heat is taken from the air to allow evaporation of moisture, then it is shown by a line drawn along the wet-bulb line (adiabatic humidification). Then the dry-bulb temperature on the bottom scale and the moisture content on the vertical scale both change.

Cooling is depicted by a horizontal line drawn from right to left of the chart. This is the converse of heating, so the dry-bulb temperature decreases along the bottom scale with no change of moisture content on the vertical scale.

Dehumidification is shown by a vertical line from top to bottom of the chart. The moisture content decreases on the right-hand vertical scale.

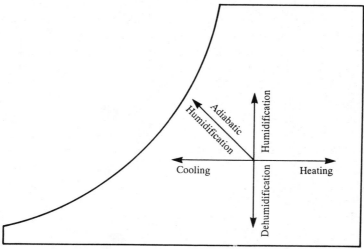

Figure 3.9 *Psychrometric processes*

Air mixing can be shown by drawing a line between two conditions of the airstreams being mixed and calculating a point in relation to the proportions of the mass flow rates of the two airstreams (Figure 3.10).

As air passes through different components of an air conditioning system its condition changes according to the component function and these changes can be depicted on the chart. For example, when air passes through a heater battery or coil, sensible heat is added and its temperature increases. This will be shown as a horizontal line from left to right on the chart. After passing through the coil the air could then go through an air washer or humidifier and this process could be indicated by a line drawn diagonally along the wet-bulb line. The air could then be reheated, shown as another horizontal line on the chart. To complete a cycle the line should return to the original condition after being altered by the various parts of the system and the conditioned space.

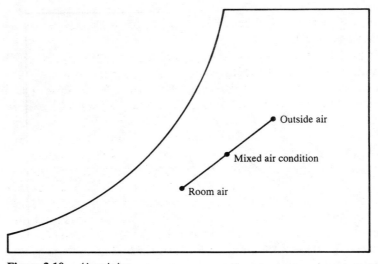

Figure 3.10 *Air mixing*

Most humidifiers used in packaged air conditioners are steam humidifiers, where the water is heated to form a vapour which is entrained in the airstream.

Where moisture is added by a spray, heat is taken from the air to promote evaporation into the airstream, sometimes called evaporative cooling. Some so-called air conditioners operate on this principle, but although they do cool they also add moisture to the air and do not therefore provide comfort conditions in countries having high ambient humidity such as the UK. Their use should be confined to areas having low humidity.

Often there is no requirement for cooling, but a need to reduce the humidity in such applications as storage areas or swimming pools. In these situations, a dehumidifier provides the solution. This is a unit which is similar to an air conditioner but is designed to remove the maximum moisture by cooling the air far below its dew-point. To avoid blowing cold air into the space, the air leaving the evaporator or cooling coil is reheated by the condenser or heat-rejecting coil. Thus the air enters the space much drier and slightly heated. On the chart this is depicted as a line sloping down from right to left from the room condition, then going horizontally from left to right beyond the room dry-bulb temperature depicting the reheat.

In practice, the process of cooling and dehumidification is not a sloping line. Some of the air entering the cooling coil is sensibly cooled by contact with the coil fins and the process goes horizontally from right to left until it reaches the saturation line (100% RH) which is its dew-point. The air is further cooled because of the low evaporation temperature and gives off moisture following the saturation curve. A portion of the air bypasses the fins on its passage through the coil and remains at room temperature. Thus there is a mixture condition of air leaving the coil of that cooled and that remaining at room condition. This is in proportion to the coil efficiency: the more rows the greater the efficiency. This mixed condition can be shown on a line drawn between the theoretical point on the saturation line and the room air condition.

Sensible heat ratio

The slope of the line indicates the proportion of sensible heat causing a change in dry-bulb temperature, indicated by a horizontal line, and the latent heat which causes a change in moisture content in the air, indicated by a vertical line. Where no latent heat needs to be removed, such as in heating, then the line would be horizontal; but as the latent heat to be removed increases, the slope of the line becomes steeper. This proportion is known as the sensible heat ratio and can be determined by dividing the sensible heat by the total heat. The slope of the sensible heat ratio can be taken from the small protractor in the top left-hand corner of the psychrometric chart (see Figure 3.8) which can then be transferred to the main part of the chart.

Air conditioning system changes

The way that the condition of the air is modified by passing through the system and the air conditioned room can be graphically depicted on the psychrometric chart. Each point plotted on the chart represents a specific set of conditions of the air.

The initial design conditions specify the internal and external temperature conditions. The load estimate determines the sensible and latent cooling loads, thus giving the room sensible heat ratio.

A point can be made on the chart (Figure 3.11) which represents the conditions of the inside or room air. A similar point can be plotted representing the outside air condition. A line can be drawn from the room air condition sloping down from right to left at an angle determined by the room sensible heat ratio. Any point along this line would give a change to the room condition which would remove the right proportion of sensible and latent heat from the room. A point selected near the room condition, giving a low temperature difference room air and supply or entering air, would require a greater amount of air to be circulated through the system to remove the calculated heat load than if the point was selected

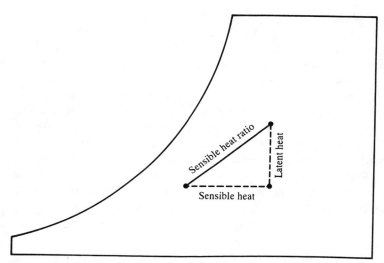

Figure 3.11 *Sensible heat ratio*

further away, giving a greater temperature difference. However, if this temperature difference is too great the temperature of the air entering the room will be low, possibly causing cold draughts. Ideally the difference should be selected to give the lowest air volume which can be circulated through the system without causing draughts.

To an extent in packaged air conditioning, the temperature of the air leaving the unit and hence the supply air is determined by the manufacturer when designing the unit, which limits the choice of temperature differences. For non-ducted single-packaged or splits systems the temperature difference will usually be between 14°C and 16°C, whereas the difference for ducted systems will be about 12°C. Using the appropriate temperature difference, a point can be plotted on the sloping line representing the supply air temperature.

If no additional ventilation is required because that from infiltration is sufficient, then room air and the outside air are not mixed before entering the unit. In these circumstances the sloping line will not only represent the change in conditions of the air as it passes through the room where it will be heated and humidified by the room load, but also the change when passing through the unit where it will be cooled and dehumidified by the unit.

If additional ventilation air is needed and it is intended to provide this air through the equipment rather than directly into the room, then the ratio through the equipment will be different from that of the room. The mixed air will be a point on a line drawn between the outside air condition and the room air condition. The equipment sensible heat ratio will be a condition represented by a line drawn from the mixed air condition to the supply air condition. This line will have a slope which is steeper than the room sensible heat ratio and hence will represent a lower ratio than that of the room.

This is an important characteristic when selecting equipment, ensuring that the proportions of sensible and latent heat which are removed from the room at one condition are removed from the circulating air by the unit at different conditions.

The load caused by solar radiation and internal heat gains and other heat sources can create unacceptable temperature levels within the space.

The air conditioning system changes plotted in Figure 3.12 are for a unit using mechanical refrigeration to reduce those levels.

Cooling by ventilation

Mechanical ventilation, using outside air, can alleviate the situation but it can never bring the internal temperature to a level at or below the outside temperature. The conditions obtained are often insufficient to provide comfortable temperature levels and can also lead to excessive air changes and air movement without benefit to the occupants. Also, a ventilation system does not control the build-up of humidity within a building.

Mechanical ventilation therefore, can be effective where high numbers of air changes do not matter, such as in factories dealing

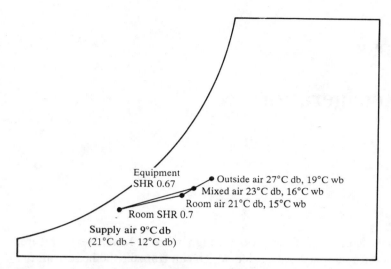

Figure 3.12 *Changes through air conditioning system*

with certain processes where the comfort of the operators is of less importance, or where no one will be present, for example when it is used for precooling overnight during the air conditioning plant shut down. It can also be effective in swimming pool applications, especially with heat recovery systems.

FOUR

Refrigeration

Heat can only flow from a higher to a lower temperature; therefore a system is needed which can transfer the heat from the lower temperature in the room to the higher outside temperature and also be capable of removing unwanted moisture from the air. Such a system is a refrigeration system which, like a water pump pumping water uphill, pumps heat 'uphill' from a lower temperature to a higher temperature.

The two most commonly used basic refrigeration systems used in air conditioning are the vapour compression cycle (used by over 90% of all units) and to a lesser extent the absorption cycle. Other types of refrigeration have been applied to air conditioning but are outside the standard range of equipment.

Vapour compression cycle

The vapour compression cycle (Figure 4.1) employs two basic thermodynamic characteristics of fluids to transfer heat from a lower temperature to a higher temperature. The first characteristic is that as the fluid's saturation pressure increases, so does its saturation temperature and vice versa. The second characteristic is that as a fluid changes state, i.e. from liquid to gas, it absorbs much more heat than when it simply changes temperature. This

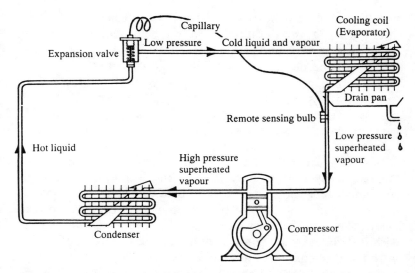

Figure 4.1 *Vapour compression refrigeration cycle (By courtesy of the Electricity Association)*

second characteristic is called latent heat of vaporization, where the change of state is from liquid to gas.

Refrigerants

Water has been used as a refrigerant and has additional characteristics that would make it ideal, except for one major disadvantage. It does not change state at normal ambient temperatures or atmospheric pressures. As an example, water boils at 100°C at sea level yet will boil at 90°C on top of a mountain where the atmospheric pressure is lower.

Some other standard substances such as ethyl chloride, sulphur dioxide and ammonia were once used but had disadvantages. In the 1930s a range of fluids known as halogenated hydrocarbons were developed as refrigerant fluids which changed state at the normal operational temperatures and which were required to be non-flammable, non-corrosive and non-toxic. The two most commonly used in air conditioning were dichlorodifluoromethane, designated as R12 and known as a CFC, and monochlorodifluoromethane, designated as R22 and known as an HCFC. Later another refrigerant was used, designated R502, which is an azeotropic mixture of R22 and R115. R502 was generally treated as a special refrigerant by the air conditioning industry and was never purchased in sufficient quantities to match the price of other refrigerants.

In recent years CFCs have been proved to cause a depletion of the ozone layer when released into the atmosphere, so their continued long-term use is now banned. HCFCs have less depletion potential and are still being used in packaged equipment. The manufacturers are currently developing alternatives with little or no depletion potential and these will gradually be introduced in conjunction with new equipment.

Although the hydrocarbon refrigerants are not normally toxic or corrosive, under certain circumstances they can be both. Burning refrigerant, for example, can result in the production of a poisonous gas. This would not of course happen under normal circumstances, but care needs to be exercised if the refrigerant circuit is routed through an area where naked flames may be present, such as laboratories.

Should there be a leak even with no flame present there is still a need for care. At certain concentration levels, the refrigerant gas, because that is what it will be at atmospheric pressure, can have a narcotic effect on the occupants of the space where the leak occurs and can also be cardiac sensitive. There are certain recommended levels of concentration for each refrigerant that should not be exceeded in the case of a major leak. These are given in BS 4434:1989 as the level that can be allowed in the 'smallest human occupied space'. These concentration levels have an effect of limiting the refrigerant charge of a system where the pipes go through or the equipment is in the occupied space. For most single-zone packaged air conditioning applications the charge will give concentrations below those recommended. However, in the larger modern and complex systems this matter may need careful consideration.

Throughout the world there is concern that refrigerant released into the atmosphere will cause damage to the ozone layer and although HCFCs are still permitted, they may be banned in some countries in the near future. Whatever the situation there will be a need to ensure that new or existing installations do not discharge refrigerant into the environment. There will be in the future an increasing number of recommendations and even laws to minimize such discharge. For example, pressure relief devices in systems will be required to vent into another part of the system rather than to atmosphere, and more isolating valves or similar devices to minimize loss when the equipment is being maintained or repaired. Such recommendations will be embodied in the European Standard EN 378 dealing with refrigeration safety and environmental protection.

Each refrigerant has different properties, particularly with respect to pressure, temperature and enthalpy, and refrigerant selection is to suit the system operating requirements for different processes. These properties can be depicted on a pressure enthalpy diagram or chart which is particular for each refrigerant type (Figure 4.2).

Direct expansion unit

Whatever refrigerant is used the refrigeration cycle is the same, and for packaged equipment most are of the direct expansion (DX) type. The term 'direct expansion' is used to signify direct heat transfer from the medium to be cooled to the refrigerant. In this case the medium to be cooled is air from the room which is passed directly over the refrigerant-filled cooling coil, known as the evaporator. As the air passes over the coil, heat is absorbed by the low-pressure, low-temperature liquid refrigerant, causing it to boil or evaporate into a low-pressure, low-temperature gas. This gas is drawn from the evaporator by means of a compressor, which compresses the gas to a higher pressure and hence a higher

Packaged Air Conditioning

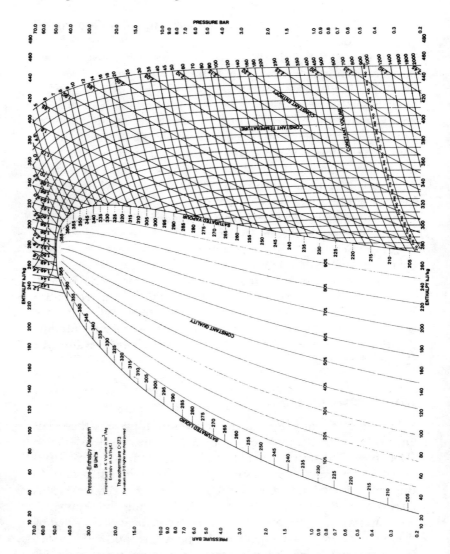

Figure 4.2 *Pressure enthalpy diagram (By courtesy of ICI Chemicals and Polymers)*

temperature. This high pressure gas is delivered out of the compressor to another coil known as a condenser. Ambient air is blown across the condenser which cools the high-pressure, high-temperature refrigerant gas until it condenses back to a liquid. The condensed refrigerant is then returned to the evaporator via an expansion device, which changes the refrigerant back to a low-pressure, low-temperature liquid to start the cycle again. The cycle is thus carried out in a sealed system under pressure, from which no leakage should occur.

For DX units, usually both the evaporator and condenser coils are made from finned tube, although if the unit is water cooled a shell and pipe coil condenser is used. The compressors used for most packaged units are hermetic which means the compressor and its motor are sealed in an airtight container. This enables the motor to be cooled by the refrigerant and avoids problems of leaks past drive seals. The compressor may be either reciprocating, rotary or scroll, depending on the maker and the application.

The reciprocating compressor is a positive displacement unit which compresses by means of pistons moving up and down in a cylinder, in a manner similar to an automobile engine. The refrigerant/oil mixture is drawn into the cylinder on the piston's down-stroke and compressed on its upward stroke. Until the early 1980s this was the main type of compressor used in packaged air conditioning equipment. The main disadvantage was that there was a limit to the pressure difference between suction and discharge which affected its use for heat pumps and also limited the length of refrigerant pipe. Care must be taken in the design to ensure that no liquid refrigerant is drawn into the cylinder, otherwise severe damage can be caused to the compressor.

A development which overcame many of these problems was the rotary vane compressor (Figure 4.3). This is also a positive displacement compressor where the compression is obtained by the action of an eccentric cam against a vane within a cylinder-like housing. Such a compressor can operate against high-pressure differentials, is quieter and smaller than the equivalent reciprocating compressors and can handle liquid in the suction without

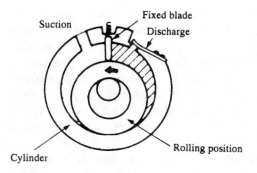

Figure 4.3 *Rotary vane compressor (By courtesy of Daikin Europe NV)*

damage to the unit. This compressor has virtually replaced the reciprocating compressor in all the smaller packaged air conditioning applications.

For larger applications in the packaged air conditioning market the scroll compressor (Figure 4.4) is replacing the reciprocating compressor. This compressor has a fixed raised involute or spiral, called a scroll, into which fits a similar grooved scroll on an eccentric rotating plate, bringing the scrolls together and causing the refrigerant between them to be compressed. Like the rotary compressor it can operate at higher pressures and deal with liquid in the suction.

For very small units the expansion device may be a length of small-bore capillary tube, whereas on larger units it is more likely to be a mechanically operated bellows, thermostatic expansion needle orifice valve or an electronic expansion valve.

The capillary tube is self-regulating for small amounts of refrigerant and this, combined with its simplicity, is an advantage.

The thermostatic expansion valve regulates the flow of refrigerant passing around the system by opening or closing the orifice valve by a sensor strapped on the evaporator outlet. These devices control the flow to give superheated gas at the suction side of the compressor and are the most commonly used where there is a

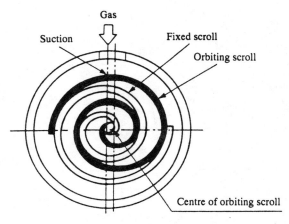

Figure 4.4 *Scroll compressor (By courtesy of Daikin Europe NV)*

single evaporator in the refrigeration cycle. The disadvantage of their use with multiple evaporator coils on one common circuit is that the expansion valve sensor has to be near the evaporator which it is to control, thus it can only control one coil, leaving the others effectively uncontrolled. More expensive, but much more effective, is the electronic expansion valve.

The pressure enthalpy diagram

The designer of refrigeration systems, as well as the engineer installing and maintaining them, needs to know the changes that occur to the refrigerant at various stages in the cycle. To do this a thermodynamic pressure enthalpy (PE) diagram (Figure 4.5) is used which gives the total heat content of the particular refrigerant at its various states of liquid, vapour and superheated gas for any given pressure within its operating range. The envelope is the dividing line between those states, on the left between liquid and vapour and on the right between vapour and superheated gas. Also shown on the chart are the absolute temperatures at the various states and pressures.

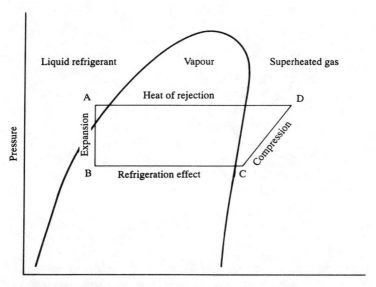

Figure 4.5 *Pressure enthalpy diagram showing cycle*

A line can be plotted from a point A, denoting a high-pressure liquid, which is the state of the refrigerant when leaving the condenser before entering the expansion valve. This line is drawn vertically down to a point B within the mixture state, depicting the point of entry as a low-pressure liquid/vapour mixture entering the evaporator. From this point a horizontal line can be drawn across the chart to a point C in the superheated gas state, representing the refrigerant picking up heat in the evaporator at constant pressure. The low-pressure gas then enters the compressor, which can be represented as a diagonal line rising from left to right up the chart to point D as the refrigerant pressure and gas temperature is increased. The line is diagonal as the enthalpy of the refrigerant also increases from the motor input to the compressor. A line can then be drawn from the high-pressure discharge from the compressor point D horizontally to the initial point A, representing the passage of the refrigerant through the condenser at constant pressure, where it changes from a high-temperature superheated gas to a mixture, then to a liquid, all at high pressure. This is an idealized representation of the refrigeration cycle by component function and does not indicate the actual pressure and enthalpy changes which also occur during the system. These occur in connecting pipework between the various components and in the practical matching of components in the cycle.

The horizontal line from the expansion valve through the evaporator represents the refrigeration effect of the evaporator; the diagonal line represents the work done by the motor-driven compressor; and the upper horizontal line from the compressor through the condenser represents the heat of condensation.

The measurement of efficiency of the refrigeration cycle is called the coefficient of performance (COP), with a subscript suffix R for the efficiency when the unit is used for cooling and a subscript suffix H for the efficiency when the heat of condensation is used for heating and the unit is known as a heat pump.

COP_R is the refrigeration effect over the work done, usually between 3 to 1 and 4 to 1 for DX packaged air conditioning units. COP_H is the heat of condensation over the work done. As the heat of condensation is the refrigeration effect plus the work done, COP_H is always 1 greater than COP_R.

These theoretical COPs for the refrigeration cycle should not be confused with the actual operating coefficient of performance of the air conditioning unit or heat pump.

In practice, the operational COP for the refrigeration cycle will depend on the operating conditions, the equipment match and the system application or efficiency. The operational COP of an air conditioning unit or heat pump, having an electric motor-driven compressor, in the cooling mode, normally quoted as an energy efficiency ratio (EER), is about 3 to 1 maximum. The COP in the heating mode should include not only the power of the compressor, but all other power using components, as well as that required for defrosting. Due to the changes in outside temperature, this can

43

vary from 4 to 1 down to 1 to 1. Generally, the seasonal COP over a winter for a heat pump is about 2.5 to 1.

Always be sure of which COP is being quoted, whether it is the compressor COP, the equipment COP or the seasonal COP.

Absorption cycle

The absorption refrigeration cycle (Figure 4.6) uses a mixture of two chemicals to provide a refrigerant/absorbent combination to transfer heat around the cycle. There are a number of different combinations, but the most common are ammonia/water used for small packaged systems and water/lithium bromide used for larger commercial or industrial applications.

Like the vapour compression cycle, the absorption cycle also operates with a pressure difference between the evaporator and condenser and has components that perform basically the same

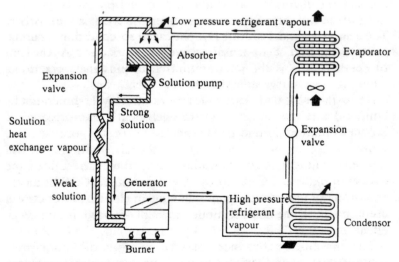

Figure 4.6 *Absorption refrigeration cycle (By courtesy of the Chartered Institution of Building Services Engineers)*

function. However, the pressure difference is not produced by a compressor but by a liquid pump and the effect of heat on the refrigerant/absorbent combination.

Liquid refrigerant passes through an evaporator where it is changed to a low-pressure vapour. The evaporator can be directly in the airstream or have a secondary medium, usually water. From the evaporator the low-pressure vapour is drawn into an absorber to mix with the absorbent. The heat in the refrigerant is passed to this mixture during the process. To increase the efficiency of the heat transfer, the outside of the absorber is cooled. The concentrated solution of refrigerant and absorbent is pumped by a high-pressure liquid pump to the generator on the high-pressure side of the cycle. The generator heats the mixture to boil off the refrigerant where it passes as a high-pressure vapour to the condenser. The refrigerant is condensed to a liquid and then reduced to low pressure by an expansion valve before returning to the evaporator to begin the cycle again. To complete the process, the absorbent left in the generator must be returned to the absorber. This absorbent passes through a heat exchanger and then through an expansion valve to be sprayed into the absorber. The heat exchanger recovers heat from the absorbent, leaving the generator to heat that entering the generator and thus making the process more efficient. The additional expansion valve is needed to allow the high-pressure absorbent to be transferred to the low-pressure absorber. Heat which is needed by the generator to boil off the refrigerant is provided either directly from a gas or oil burner or indirectly on larger units by steam or high-pressure hot water.

The advantages of an absorption machine are that it does not use ozone-depleting CFCs, although it does use other chemicals and does not require a large power supply, as the only electrical use is for pumping the solutions between the various compartments.

These advantages, however, may be outweighed by the disadvantages of size, low thermal efficiency, increased capital costs and specialized maintenance requirements.

The COP of absorption machines varies between 0.5 to 1 and 0.8 to 1 in practice, although COPs in excess of 1 to 1 are claimed for complex experimental systems. As such, the equipment requires much more heating power than a vapour compression machine. Assuming a heater efficiency of 70% against an electricity generating efficiency of 30%, the absorption machine would have to have a COP of better than 1.3 to 1 to use less prime energy than the electrically driven vapour compression machine.

FIVE

Equipment design and construction

Elements of a packaged air conditioning unit

Packaged equipment is pre-engineered and built on a production line where the manufacturer has designed the unit to work within pre-selected conditions, using standard components which includes a preset refrigeration circuit to achieve those conditions.

Although a chilled water unit, where the refrigeration cycle chills water, to be pumped around the building, is often pre-engineered, the term 'packaged air conditioning' has usually been confined to direct expansion units, where the refrigeration system cools the air to be conditioned without an intermediate medium.

Chilled water systems have tended to be central station systems where the equipment is located in main plant rooms. Packaged equipment is normally decentralized and located close to or in the area to be conditioned, although this may not be the case with some multizone systems.

Packaged equipment tends to be matched indoor and outdoor units from the same manufacturer which is how they should be applied. Designers of air conditioning sometimes avoid using packaged equipment because they feel that they have to 'bend' their designs to match the equipment available, rather than use specially engineered central plant where they can set the equipment's characteristics to match their design.

However, with the knowledge of how the designer of pre-engineered equipment sets about his task, and of the way the equipment will operate over a range of conditions to those selected by the designer, the air conditioning system engineer can often obtain a very close match to his design requirements.

Consideration should be given to how close a match needs to be made to a theoretical design when in practice that design may bear little resemblance to the way it actually will operate. If it is to be properly applied and correctly selected, then an understanding of the product and the technology behind it is vital together with the effect of changing operating conditions.

Packaged air conditioning equipment can be 'cooling only', and termed air conditioners, or the refrigeration cycle can be reversed to provide either heating or cooling and are called reverse cycle heat pumps.

All systems take heat out of the room or area to be cooled and reject that heat outside the conditioned space. The unit has a section which absorbs heat from the space and another section which rejects the heat outside the space. These sections can be within the one casing or be split between two or more casings. When the sections are split the heat-absorbing section is known as the fan coil unit or alternatively the indoor unit, whereas the heat-rejecting section is known as the condensing unit or the outdoor unit (Figure 5.1).

The heat taken out can be both sensible and latent heat, so the unit will deal with both cooling and dehumidification. It can also provide a degree of filtration and possibly 'fresh air' ventilation. Heating is by either reversing the cycle or by a supplementary heater. Packaged units designed for close control can have a steam humidifier within the casing to provide humidity control. In most comfort applications, the majority of which are in commercial situations, the need to provide winter humidity control is marginal and can be energy intensive.

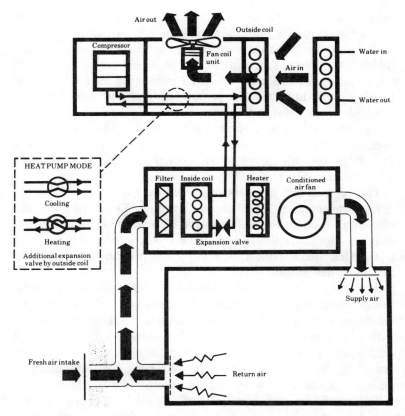

Figure 5.1 *Packaged air conditioning system (By courtesy of the Electricity Association)*

Indoor unit

The heat-absorbing section or indoor unit usually consists of a filter, coil, fan and sometimes a heater battery. The filter will normally be expanded polyurethane in the smaller room units or a series of flat glass fibre panels on the larger packaged units. These

49

are not very efficient filters and are mainly intended to ensure that dirt or lint do not get onto the finned coil and so cause a blockage. Some room unit manufacturers provide an electrostatic filter, which is capable of removing pollen or cigarette smoke from the air, as an optional accessory. Other more efficient filters can be fitted if the additional fan power required for these filters is available from the unit fan.

Care should be taken when a unit is fitted in a location of high contamination such as a kitchen or hairdressing salon. In the former grease and in the latter lacquer can quickly block the normal filters.

The cooling coil or evaporator will normally be constructed from copper tubes with aluminium fins. Depending on the efficiency required, the coil can have between two and eight rows of tubes with seven fins in each 10 mm. Below the coil is a tray or trough designed to collect the water which is condensed out of the air below dew-point.

The type of fan incorporated will depend on the size of the unit, the physical restrictions caused by the shape of the unit, the resistance to air flow through the unit and components together with any attached ducting. On larger units it will always be a centrifugal fan, normally with forward-curved blades. On smaller units the fan is usually a tangential fan, although specially designed propeller fans are used on some types of unit.

The centrifugal fan is a series of fan blades mounted on the periphery of an impeller rotating in an involute or scroll-shaped casing. The air enters the fan through an inlet(s) on the axis of the impeller and is drawn through 90° to be discharged from the outlet which is in line with the fan's rotation. The blades on the impeller can either curve forward in the direction of rotation or backward, depending on the design.

Forward-curved fans, because of their scoop effect, will handle greater volumes of air at lower pressures than backward-curved fans, although the latter are more efficient. Also backward-curved fans have non-overloading characteristics, in that the fan motor will not be overloaded in the event of an excessive increase in pressure such as a blockage. Usually the width of the impeller is less than the diameter of the fan, but a variation is the tangential fan. This is used for small indoor sections of room air conditioners where, by having a small diameter with a long length, a compact unit with good air distribution is possible.

The propeller fan has an impeller with a series of blades at an angle to the rotor, so that it draws air axially through a plate or cylinder in which the blades are mounted. Such a fan is capable of handling large quantities of air but, unless specially designed, only against very low resistance. Propeller fans tend to be noisier than centrifugal fans for similar air quantities or capacities.

The capacity of a particular fan depends on the type, shape and size of the blades and the speed of rotation. This capacity and its ability to overcome system resistance obeys the basic fan laws which affect all air movement in air conditioning systems.

For constant density the following fan laws apply:

(a) Volume is proportional to speed.
(b) Resistance is proportional to speed squared.
(c) Power is proportional to speed cubed.

Where the fan is of the same type, operating at the same speed with air at constant density, the volume is inversely proportional to the resistance; therefore, if the resistance increases the volume delivered decreases and vice versa.

From the fan laws you also can see that with a greater resistance to air `flow not only does the volume reduce but the power absorbed or required by the fan motor can be dramatically increased.

The heater battery for an air conditioner, if required, is normally a sheathed element electric heater, but can be a low-pressure hot water coil served from the heating system. There are some larger units which incorporate a gas-fired warm air heater to provide any heating requirement. If the unit is a reverse cycle heat pump a supplementary heater may still be incorporated to temper the air during defrosting periods or to add to the heat from the refrigeration system.

Outdoor unit

The heat-rejecting section or outdoor unit will normally contain the compressor, a heat-rejecting coil, a fan and refrigeration controls and safety devices.

The compressor is the heart of the air conditioner. All the early packaged units and many of those in use today have reciprocating compressors. However, more recently units have been introduced using either rotary or scroll compressors which have enabled their advantages to extend the range of applications and to reduce both size and noise levels.

The heat-rejecting coil is known as the condenser and is similar in construction to the evaporator coil, but as there is no concern about latent heat it will usually consist of less rows. However, the construction may be changed if the surrounding air is liable to corrode the aluminium fins, as is often the case with installations near the sea, where the salt in the air can cause severe corrosion.

The fan in the rejecting section, which is known as a condenser fan, is usually a propeller fan as it can deal with large air quantities against a low external resistance. However, if air to the condenser needs to be ducted and therefore operate against a high resistance, or if there is need to reduce the noise from the unit, then a centrifugal fan is used.

Apart from these major components there are other items such as thermostats which are included to deal with the control of the unit or temperature- or pressure-limiting devices to ensure its proper operation. These are considered in a separate chapter.

If the unit is to operate as a reverse cycle heat pump, rather than an air conditioner, then additional components are required such as a reversing valve. If an air conditioning unit could be physically turned around, the heat could be taken from the outside air by the evaporator and rejected inside by the condenser. Some units are designed to operate in this mode only and are called 'heating only heat pumps'. However, especially for air conditioning, it is more convenient to have a unit which can operate in either mode without the need to physically move the unit. This is achieved by

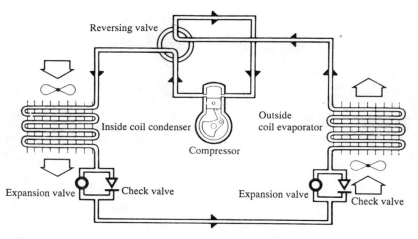

Figure 5.2 *Vapour compression heat pump – heating cycle (By courtesy of the Electricity Association)*

altering or reversing the flow of the refrigerant. Actually it is only the flow to the two coils which is altered because the direction of flow must always be the same through the compressor.

Figure 5.2 shows the refrigeration cycle when the unit is operating in the heating mode. The discharge from the compressor is directed by the reversing valve to the inside coil. The refrigerant flow goes through the check valve, bypassing the expansion device on the indoor coil, then flowing through the expansion device on the outdoor coil, through the outdoor coil and back to the compressor via the reversing valve.

Figure 5.3 shows the refrigeration cycle when the unit is operating in the cooling mode. The discharge from the compressor is now directed at the outdoor coil by the reversing valve and the outdoor coil expansion device is bypassed.

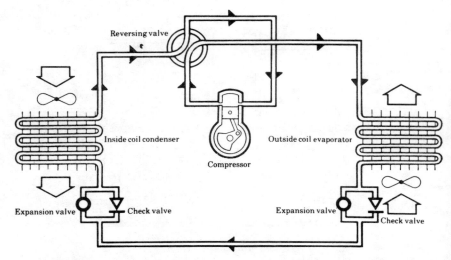

Figure 5.3 *Vapour compression heat pump – cooling cycle (By courtesy of the Electricity Association)*

Reduction of noise level

The major components in a packaged unit which create noise are the compressor and the fan. The design of modern compressors have significantly reduced their noise and it is usually the fans which now cause the most problems.

The noise which will be emitted by a fan is through either the case or the ducting system. The noise is related to the tip speed of the fan blades. This speed depends on the rotational speed of the fan and its diameter. Therefore, the faster the fan speed, the greater the noise.

The noise level can be measured as a sound power level in watts and represents the total rate of noise. As the value is absolute, it can express the noise value regardless of application or situation. However, what the designer has to consider is the perceived sound heard by the human ear or that caused by vibration. This is expressed as a sound pressure level and depends on both source and the surroundings. Because of this, sound pressure levels must give an indication of the point from the source at which it was measured, together with other relevant information about the surroundings.

The decibel is the unit used for measuring sound. Noise meters are weighted so that they measure the decibel level in slightly different ways. The weighting normally used to indicate the background noise is known as the 'A' scale written as dBA and gives the sound level response nearest to that of the human ear. Generally, the lower the level the less noise, but level is not the only consideration, since the perception of noise varies with frequency as well as level. At high frequencies the decibel level needs to be lower than that acceptable at low frequencies to give similar perceived noise. High-frequency noise can usually be attenuated without too much difficulty, but this is not the case with low-frequency noise which requires special consideration. An octave band analysis will assist in comparing the perceived noise level at different frequencies and can be given in two ways – as a noise rating (NR) or as a noise criterion (NC), depending on the octave bands chosen.

The following levels indicate the level of noise in dBA which can be expected as background noise in various applications:

Recording rooms, churches, private bedrooms	25
Private living rooms, hotel bedrooms, libraries	30
Public hotel rooms, conference rooms, private offices	35
General offices, receptions, restaurants, stores	40
Computer suites, supermarkets, kitchens, canteens	45
Machine shops and similar industrial situations	above 50

Where possible, the equipment selected should have a noise level less than the background of the application. The outside unit of the equipment could easily affect and annoy other people even if it is not disturbing those occupying the space. There are laws and regulations which limit the nuisance that you may cause by noisy equipment. The local environmental health officer deals with these matters and may be able to advise on the acceptable levels. He also has the powers to stop the use of equipment which exceeds the levels, so it is important to consider the effect of the fan or equipment in a particular location. This consideration may be even more important for 24-hour use or when considering heat pumps which can operate in the early hours of a winter's morning for preheating, as a noisy area during the normal working day could become very quiet in the evening or at night. A system which is acceptable during the day could annoy people at other times.

A reduction in noise by closing a window in the air conditioned building will not be effective if the window in an adjoining property is opened for ventilation. For example, a unit having a noise level of say 45 dBA some 2 metres away from a property will be attenuated to about 35 dBA by a closed window to a bedroom. This could be acceptable, but if the window is opened the level will rise towards 45 dBA which becomes entirely unacceptable.

In a commercial building there could be residential accommodation where noise levels acceptable for businesses are not so to those in the living accommodation. Generally, noise levels are measured at the boundary, and distance can help to attenuate the noise.

Care always needs to be taken to minimize the creation and transmission of noise and vibration. Sometimes the problems, especially if the noise is of a high frequency, can be solved by erecting sound-absorbing barriers around the equipment or by specially constructed attenuators fitted to the equipment. However, the cheapest solution is to avoid the problem by the proper selection of equipment. For example, the use of a centrifugal fan and ducting the condenser air to and from a unit in an enclosed plant room will reduce the possibility of exterior noise. This avoids a problem, and is preferable to using expensive attenuation to minimize the noise caused by an outside condensing unit with a propeller fan, which may initially seem to be a cheap alternative.

Noise, especially from fans, together with overflowing condensate trays and blocked filters, are the major areas of complaint from users or neighbours of air conditioning systems.

Designing the appropriate unit

To understand how packaged air conditioning equipment operates and its possible limitations it is necessary to have an understanding of how an equipment designer goes about producing a unit for a given application. Despite the number of computer programs available dealing with heat transfer through coils, the characteristics of compressors, etc., the design of a unit still involves a certain amount of art as well as science. The basic tools used by the designer are the refrigerant pressure enthalpy (PE) diagram, the psychrometric chart, a knowledge of the characteristics of a range of compressors and the heat transfer characteristics of finned copper tube coils.

Information regarding the compressor enables the designer to select a unit which will operate in the required pressure range, with the appropriate refrigerant passing the necessary mass flow of refrigerant to transfer the heat from the low-pressure evaporator to the high-pressure condenser.

The characteristics of heat transfer of the coils, including the coil effectiveness, help the designer to select the right size of coil to absorb heat as an evaporator and to reject heat as a condenser.

The balance of the system is achieved by matching the compressor duty to the expansion device duty.

Assessing the conditions

Having all this information to hand, where does the equipment designer start? First he must decide on the application or range of applications proposed for the equipment, for example whether it is to be used in a private office, or a computer room requiring a high sensible heat ratio, or a crowded shop or restaurant needing a low sensible heat ratio. Then he must assess what the inside and outside environmental conditions will be and whether they are to be based on conditions for a specific location, or more general worldwide conditions, or extreme tropical conditions. Having determined the conditions he can plot them on a psychrometric

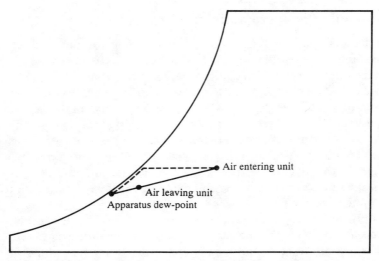

Figure 5.4   *Psychrometric chart showing apparatus dew-point*

chart. They will be the selected external dry- and wet-bulb temperatures and the internal dry- and wet-bulb temperatures.

A sloping line, equivalent to the sensible heat ratio of the selected application, drawn from the point representing the inside conditions until it crosses the saturation line, that is the line of 100% humidity, will give a point known as the apparatus dew-point (ADP), as shown in Figure 5.4. This is the theoretical temperature at which the evaporator coil would operate if the coil was 100% efficient. It is not the temperature of the air leaving the coil; this is determined by the coil effectiveness or its 'bypass factor' (Table 5.1). This bypass factor is the ratio of the amount of entering air which is affected by the cold coil when passing close to the fins, compared with that less affected because it passes in a stream between the fins and is therefore not cooled.

As air goes through a row of a coil, some of the air is cooled down to the coil temperature and some is unaffected. These two conditions of air are mixed when leaving the row, giving a

Table 5.1 Typical air conditioning coil bypass factors (By courtesy of Carrier Air Conditioning)*

No. of rows of coil	Factor
2	0.31
3	0.18
4	0.10
5	0.06
6	0.03

*The bypass factors are also affected by the velocity of the air passing over the coil and the pitch of the fins. The faster the air velocity, the greater the bypass factor. Also the closer the fin pitch, the smaller the bypass factor

condition between the two. A similar process occurs as it enters the next row, and so on. Thus, the more rows to a coil, the more effective it is and the greater the amount of heat and moisture that is removed. The speed of the air through the coil will also affect its effectiveness. The higher the speed, the less effective the coil because it has less chance to give up heat; alternatively, the lower the speed, the more effective is the coil.

The designer now knows, by using the psychrometric chart, the condition of the air entering the evaporator, the theoretical evaporator operating temperature and, from the bypass factor of the coil selected, the probable condition of the air leaving the coil. The theoretical operating temperature and the refrigerant selected for the application will, with the aid of a PE diagram, determine the evaporator pressure or the low-pressure side of the refrigerant cycle.

To determine the amount of refrigerant which needs to be pumped around the system to transfer the heat, the designer needs to select a total capacity for his unit. This will be the total amount of heat, sensible and latent, that the unit will reject from the conditioned space. Knowing the amount of heat to be absorbed by the evaporator and then rejected by the condenser, the designer can determine, by calculation, the amount of refrigerant mass flow needed. At the pressure conditions selected for the evaporator, the volume of refrigerant equivalent to the mass flow can be determined and this will need to be handled by the compressor. However, before the compressor and its motor can be sized, details of the condenser conditions, especially the condensing pressure, need to be determined. The designer has selected the external conditions that the unit is to operate, and to enable reasonable heat transfer from the condenser to the ambient a temperature difference of about 10°C can be selected. This then gives the condensing temperature of the refrigerant and from the PE diagram the appropriate condensing pressure.

With the information on the evaporating pressure, the condensing pressure and the volume of refrigerant to be handled, the size of the compressor and compressor motor can be determined from the data on the compressor considered. As the compressor is usually hermetic, the heat from the motor is also transferred to the refrigerant. Thus the heat to be rejected from the condenser is the heat absorbed by the evaporator and the heat of the compressor motor. The condenser coil is sized by using the information on the heat to be rejected, the temperature difference over the coil and the volume of air passing through the coil.

The fan sizes are determined by the air volumes required to transfer the heat from the air to the evaporator and from the condenser to the ambient, and any resistance to the air flow caused by the coils, filters, etc., and also with an allowance for any external ducting if designed for use with ducting.

Changes in conditions

The foregoing description is a rather simplistic account of how a unit is designed, but provides a basic understanding of the principles involved.

From this design information it is obvious that a change in any of the conditions can have a fundamental effect on the performance of the unit. For example, an increase in the air flow across the coils will have the effect of increasing the bypass factor and decreasing the coil effectiveness, but nevertheless increasing the total heat transfer of the coil.

If the ambient temperature is below that allowed by the designer, the heat transfer is greater and more heat is removed by the condenser which in turn will affect the whole refrigeration system. A reduction in load onto the evaporator will have the effect of reducing the evaporating temperature and pressure if the reduction is not properly controlled. In fact, any significant variation will change the unit's characteristics, especially in relation to its capacity and the proportion of sensible and latent heat extracted. When properly designed and used, the limiting controls in the unit will avoid it going outside its operating range by shutting the unit off when the limits are exceeded.

If a unit, which has been designed for an overseas comfort market where high internal and external temperatures with a low sensible heat ratio apply, is used in an application in the UK requiring a high sensible heat ratio such as a control room, then the design conditions will not be met. Either insufficient sensible heat will be removed and the air temperature rise or too much latent heat will be removed, making the relative humidity drop, and requiring the use of a humidifier to maintain the right balance.

SIX

Equipment types and conditions

Types of unit

Packaged air conditioning equipment can be divided into three types: self-contained, free-standing with remote condensers or split-packaged systems.

Self-contained or single-packaged unit

The single-packaged unit (Figure 6.1) has all the major components previously described, with the possible exception of the thermostat controls in one casing. The simplest single-packaged unit is the portable unit named the single-duct unit. With this unit the air for the condenser is taken from the conditioned space, but the hot discharge air is ducted away from the unit to outside by a single flexible duct. The condenser can be air cooled or water can be passed over the coil to make it an evaporative condenser. This should not be confused with the evaporative cooler referred to earlier in Chapter 3. The evaporative condenser is more efficient at heat rejection than the air cooled condenser, but the water container needs constant refilling and if it becomes empty it automatically turns off the unit. There is also a version which can operate in either mode.

In addition, there are small portable units based on the window

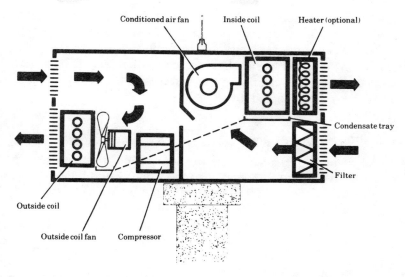

Figure 6.1 *Single-packaged unit (By courtesy of the Electricity Association)*

unit or the split-packaged unit which operate on the same principles, described in detail later in this chapter.

The most common self-contained unit throughout the world, the cheapest and least sophisticated true air conditioner, is the window unit. As its name implies it is meant to be installed in a window on the sill or over a shop doorway with the heat-rejecting side projecting outside. It is available from capacities of 1 kW up to 7 kW. The advantages are that it is cheap and simple to install. The disadvantages are noise, an unsightly external appearance and poor distribution of the inside air.

The window unit can often be mounted in a partition of an office or store within a factory or warehouse to cool the space, allowing the condenser heat to discharge into the much larger volume of the factory or warehouse. The extra heat has little effect on the larger space and the installation is a cheap way of providing comfort where conditions are bad.

Packaged Air Conditioning

They have also been used as 'spot coolers', directing a stream of cold dehumidified air to give some degree of comfort. In this instance the heat is rejected into the same space and may make conditions in other parts of the space intolerable. If, however, there are very high ceilings in the space, then the heat could rise away from the occupied area.

Console units

Another single packaged unit is the floor-mounted or console air conditioner (Figure 6.2), similar to the window unit but with the heat rejection part built into the wall of the building and covered with a decorative grille on the outside. Sizes of this unit are between 2 kW and 6 kW. The advantages are similar to the

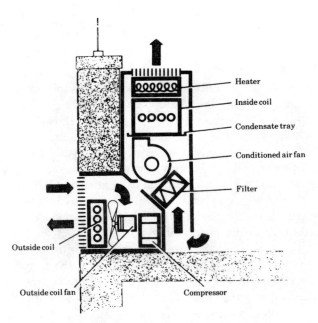

Figure 6.2 *Console unit (By courtesy of the Electricity Association)*

window unit, but the disadvantages are less because the units are quieter, less unsightly and have marginally better distribution (Figure 6.3). However, an aperture of about 600 mm × 300 mm is required to enable the unit to fit in the wall.

A variation of the console unit is one where the various components are mounted on separate chassis within the one casing. There is a box and grille for mounting in the wall as the building is being constructed. A fan coil chassis consisting of a heater battery, filter and fans is used, with an inside casing incorporating an appropriate diffuser to make the unit into a fan convector heater. Finally, a cooling chassis consisting of an evaporator, compressor, condenser and condenser fan makes the convector into an air conditioner. This enables the building services designer initially to specify only those parts which are required by the building use if air conditioning is not to be considered for the original contract.

Figure 6.3   *Console unit (By courtesy of Temperature Ltd)*

Roof and ceiling-void units

The largest of the single-packaged units is the roof unit (Figure 6.4). These are factory-assembled self-contained units with capacities from about 10 kW to over 300 kW, but with the air inlet and outlet to the evaporator normally ducted.

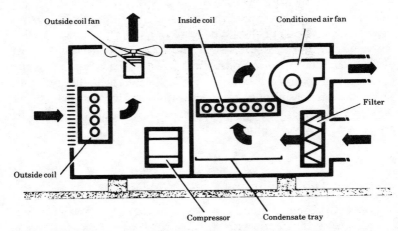

Figure 6.4 *Roof unit (By courtesy of the Electricity Association)*

A variation of the roof unit is the ceiling-void unit (Figure 6.5). This is a low-height unit with a centrifugal condenser fan enabling the condenser cooling air to be ducted to and from the unit. It can be positioned in the ceiling void inside the building or in the bulkhead over a shop doorway. In some units the evaporator and condenser sections can be split if required. These units normally have capacities from 10 kW up to 25 kW and may be used in place of a conventional unit where the noise level from a propeller fan would create problems.

66

Figure 6.5 *Ceiling-void unit (By courtesy of APV Vent-Axia Industrial Division)*

Free-standing unit with remote condensers

In this form (Figures 6.6 and 6.7) the condenser coil and fan are remote from the main unit which incorporates all the other components including the compressor. The main advantage of this unit is that it handles longer refrigerant pipe-runs with the compressor in the inside section, which may also result in lower installation cost. The condenser fans are remote, away from the conditioned space. This unit is mainly used for conditioning computer rooms, control rooms, plant rooms and similar applications. The size of units ranges from 10 kW to 70 kW.

Split-system or split-packaged unit

The main category of packaged air conditioning unit is the split system or split-packaged unit (Figure 6.8). With this unit the components are grouped into two casings, one containing the filter, fan, evaporator coil and expansion valve, known as the fan

Packaged Air Conditioning

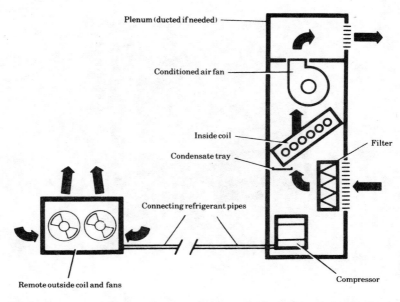

Plenum (ducted if needed)

Conditioned air fan

Inside coil

Condensate tray

Filter

Connecting refrigerant pipes

Remote outside coil and fans

Compressor

Figure 6.6 *Free-standing unit (By courtesy of the Electricity Association)*

Figure 6.7 *Free-standing unit (By courtesy of Airedale International Air Conditioning Ltd)*

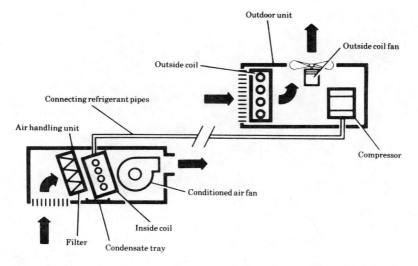

Figure 6.8 *Split-packaged unit (By courtesy of the Electricity Association)*

coil or air-handling unit, and the other containing the condenser coil and fan with the compressor, known as the condensing unit. These units range in size dependent upon the style.

The advantages of the split system are those of increased flexibility for the design and installation, because no large holes or openings are required, greater security and the capability of reducing background noise in the conditioned space.

Most units are designated split systems having a single indoor unit where one fan coil unit is matched to one outdoor unit. With modern technology there can now be split systems which have multiple indoor units to a single outdoor unit without the inherent disadvantages at one time associated with such systems. These systems are described in a later chapter dealing with systems suitable for larger multizone applications.

At the bottom of the range of split systems, with capacities from 2 kW to 7 kW, are the room units where the fan coil unit can either be wall mounted at low or high level (Figure 6.9) or ceiling

Figure 6.9   *High side-wall indoor unit (By courtesy of IMI Air Conditioning Ltd)*

Figure 6.10   *Wall-mounted condensing unit (By courtesy of IMI Air Conditioning Ltd)*

mounted with the outdoor or condensing unit mounted outside where convenient. Next are the larger units where the fan coil unit can either be vertical free-standing or horizontally mounted. These have capacities from 10 kW to 100 kW. Finally the latest version of the split system has an indoor or fan coil unit called a cassette unit (Figure 6.11). These units range from 10 kW to 20 kW capacities. They were originally intended to be mounted within a false ceiling, although many are now hung from the ceiling. The return air from the space is taken into a grille in the centre of the unit through a filter and fan to pass through an evaporator coil on one, two, three or four sides of the unit. The conditioned air is then discharged through diffusers in the unit just below the ceiling. This forms a compact air-handling unit where the discharged air can blow one, two, three or four ways according to the needs of the location.

The distance between the condensing set (Figure 6.12) and the fan coil is limited by the restrictions of the refrigerating cycle, but with a modern compressor this distance can be up to 100 m total travel, although distances above 30 m should be considered the practical limit unless special care is take during installation.

Figure 6.11 *Cassette indoor unit (By courtesy of Toshiba UK Ltd)*

Figure 6.12   *Roof-mounted condenser (By courtesy of Daikin Europe NV)*

Selection of unit

Manufacturers' data

When considering the selection of a suitable unit, a number of factors need to be taken into account. These factors relate to both the application and the unit. Where possible, the manufacturer should be requested to make the selection based on project information provided by the building services engineer. The manufacturer has all the relevant information about the design of the units, as previously discussed, and can therefore provide capacities of the unit when operating against conditions different from those given in the general technical data. However, because the original manufacturer is often overseas, or has a policy to deal only through distributors, asking the manufacturer or his distributor to select from the building design data is sometimes impossible. Where this is the situation, insist on detailed technical information covering the inside and outside conditions selected for the project.

Because the packaged air conditioning market is dominated by overseas manufacturers the data presented in general and technical brochures is often not relevant to UK conditions. Even UK manufacturers, owing to the needs of competition, will give much of their information based on these overseas conditions. For comfort room air conditioners the information is usually given for two conditions.

Overseas and UK conditions

The first, condition A, represents the typical conditions applying in the USA and similar areas which are 35°C db and 24°C wb outside, with a room temperature of 27°C db and 19°C wb. The second, condition B, represents the typical conditions of tropical climates and are 46°C db and 24°C wb outside and a room condition of 29°C db and 19°C wb.

These conditions can be quoted as those of the International Standards Organisation (ISO), British Standards (BSI) or more often the American Air Conditioning and Refrigeration Institute (ARI). The BSI also have an additional condition C, representing the conditions in the UK and other areas of northern Europe. These conditions are outside 27°C db and 19°C wb, with a room temperature of 21°C db and 15°C wb. It is hoped in the future that this condition will be included in the ISO Standards.

As far as European Standards (CEN) are concerned, all these conditions have been included in the latest drafts, together with additional requirements for specialized use such as close control or control cabinets and single-duct units or spot coolers. When these drafts are published as Standards, if compliance is quoted it will be mandatory for the manufacturers to rate the units at both conditions A and C. There are also conditions for the winter application of reverse-cycle heat pumps.

Information requirements

To select a unit the following information will need to be specified:

1 The design heating and cooling loads with both sensible heat and latent heat for the cooling loads.
2 The design external and internal temperature conditions.
3 The condition of air entering the unit or the evaporator coil (where outside air is mixed with room air).
4 The probable extreme operating conditions; for example, will the unit be cooling during the winter thereby requiring low ambient control.
5 Where the plant is to be located, and area of distribution.
6 Any special requirements such as special filtration, corrosive atmosphere, etc.
7 Acceptable internal and external noise levels.

This information can be used to select a unit from the manufacturer's data.

The product data will be the results of the manufacturer's own tests or those carried out by a third party.

This data will need to include at least the following:

(a) Heating and cooling capacities at design conditions or a performance curve giving a range of capacities.
(b) The sensible heat ratio of the equipment at design conditions or a curve which will enable an adjustment to be made from standard conditions.
(c) The air flow from the unit and whether the capacities given relate to high speed, normal speed or low speed.
(d) The upper and lower operating conditions of the unit in both the heating and the cooling modes.
(e) Method of temperature control and other limiting or safety controls.
(f) The limits of interconnecting refrigerant pipework.
(g) Noise rating for the units both inside and outside.
(h) Dimensions and weights of unit.

(i) For heat pumps, the method of defrost.

(j) Specifications for any ancillary equipment.

The requirements of a number of EEC directives relating to air conditioning deal with energy efficiency and safety. Therefore the manufacturer should provide information which will assist the calculation of the energy efficiency and demonstrate that the equipment complies with any safety requirements.

Designers should ensure that the manufacturer has the facility to carry out the appropriate tests or can produce third-party test reports giving the data.

Any reasonable manufacturer has these facilities, so if enough designers insist on receiving the correct information, the manufacturer will produce the necessary data.

At present a lot of this information is not easily available, but until it is, enabling the designer to make a correct selection of equipment, the use of packaged air conditioning will be 'hit or miss' in comparison to component central station plant and its use will be deprecated by the 'professionals'.

SEVEN

Equipment for multizone applications

Complex systems

This book is primarily written to provide information on simple packaged air conditioning and the various chapters are set out to this purpose. There are however systems which, although they use packaged equipment, are more complex than the normal units. This book does not cover all the aspects of design, installation and so on for such systems, but much is fundamental for both simple and complex systems. The following is a brief explanation of these larger systems and in simple terms how they work.

At one time packaged air conditioning equipment, because of technical limitations, was restricted to comparatively small units dealing with single-zone applications. Should a larger or multi-zone application need air conditioning, then the usual option was to use chilled water incorporated in a central plant system.

Such systems were initially a series of fan coil units supplied with chilled water or hot water as required, or induction units where primary treated air was distributed to terminal units which induced secondary or room air through the unit to be heated or cooled by secondary coils. Other systems included dual duct systems where air from a 'hot' duct was mixed at a terminal mixing box with air from a 'cold' duct to produce air at the required condition to deal with the room load. The latest is a variable air volume (VAV) system, where the volume of the air is changed at a terminal device to match the load rather than the temperature. Each system had its proponents and advantages, as well as disadvantages, but all are capital intensive. Also a significant amount of energy can be lost in either distribution or the mixing of the chilled and warm air.

A number of developments in the design and application of packaged air conditioning equipment in recent years have enabled it to be used for larger buildings and for multizone applications.

Because it was sited outside the space to be air conditioned, and therefore larger than the conventional room or free-standing units, the roof unit was used for such multizone application, ducting a supply from it to the various areas to be conditioned. One such unit incorporated a heater, normally a direct gas-fired heater, and a series of zone dampers on each of the cold air and warm air streams, so that it operated like a dual duct system. The development of the large rooftop heat pump led to its use to provide either heating or cooling as required by the application. Unfortunately the former system tended to have high mixing losses which were not acceptable in the present trend for energy economy, while the latter, being a changeover system, could not deal with the need for different conditions in each zone.

Variable volume and temperature system

With the development of the cheap micro-electronic control, however, there has been developed an air distribution system which can deal with varying zone requirements yet can still use conventional constant volume changeover heat pumps. Such a system is called a variable volume and temperature (VVT) system (Figure 7.1).

This system uses micro-electronically controlled terminal dampers, together with a central overall control to deal with the different zones. If a resistance is created to air flow, the air volume reduces; conversely if the resistance is reduced, the volume increases. It is on this principle that the VVT system works.

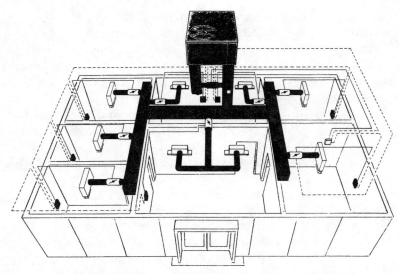

Figure 7.1   Variable volume and temperature system (By courtesy of Carrier Air Conditioning)

The system layout is the same as for any ducted constant volume air conditioning sytem, but in each zone terminal units are fitted with electronically controlled dampers in the end of the ductwork. These dampers are controlled through controllers in each zone. As the room reaches the temperature set point, the damper closes, creating an increased resistance to air flow, thus reducing the volume. The damper, by partly opening or closing, can match the volume to the load of the room. If, however, the majority of units are closed, the system fan in the air conditioning unit could be overloaded by the increase resistance. To avoid this, a controller using static pressure sensors in the variable volume part of the system operates a damper in the bypass duct between the main supply and return air streams. As the pressure changes, the bypass damper is modulated by the controller to maintain the static pressure in the variable volume part of the system. Thus as more terminal dampers close, the control ensures that the air flow and resistance are maintained within the operating range of the fan. The terminal damper controllers relay signals, indicating which zones require heating and which require cooling, so that the central controller can set the heat pump to suit the majority of zones. Thus, through the unit, the temperature of the air is controlled and through the operation of the dampers the air volume is controlled. However, technically – when the heat pump is in the cooling mode for example – it cannot provide heat to a zone requiring heating but only shut off the air supply. But in practice the system works effectively in most medium multizone applications.

Closed-loop heat reclaim system

There is a type of water-cooled air conditioner where the condenser coil and fan combination are replaced by a shell and pipe coil condenser through which water is passed to cool the refrigerant gas. There are also heat pumps which have a shell and coil in the outdoor unit which can act as an evaporator, drawing heat from a suitable source, or as a condenser, rejecting heat to the same source. These units can range from small room units up to large self-contained systems.

A number of the room units can be installed into various rooms or zones within a building, all of which can be connected to a common water circuit or loop. This application is called a closed-loop heat reclaim system (Figure 7.2).

This system uses water-to-air reverse cycle heat pumps having nominal capacities between 2 kW and 4 kW connected by a common water loop. Also connected to the loop is a suitable heat source and a means of heat rejection. The heat pumps in the cooling mode put heat into the loop, which those in the heating mode take heat out. There are many occasions when part of the building, say in a south-facing room, requires cooling, whereas another in the shade in the north requires heating. Under these situations the rejected heat into the loop provided by the units cooling is sufficient for those taking heat from the loop for heating

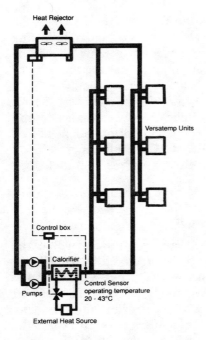

Figure 7.2 *Closed-loop heat reclaim system (By courtesy of Temperature Ltd)*

and thus the building load is in balance. When most of the units need heat and the temperature of the common loop drops, the heat source adds heat. Conversely, should most units be cooling and the loop temperature rise, the heat-rejection unit will remove the heat and maintain the balance. Throughout most of the year the system requires very little heat added or rejected. The heat source can be an air-to-water reverse cycle heat pump which can also be used to reject the heat.

The closed-loop heat reclaim system is really considered as a central plant system even though it uses packaged heat pumps. It can be an ideal system in a multizone application where flexibility is required, but there is insufficient space to run ducts and there is a limitation on refrigeration pipework.

Split systems with multiple indoor units

In terms of space needed for the medium of distribution, air requires the largest volume to deliver a set quantity of heat; water requires less, as only small-diamater pipework is needed compared with large ducts; but that requiring the smallest space is refrigerant. It requires about 30 times less space than water and nearly 200 times less space than air. Thus, the heat lost or gained through the wall of the distribution pipe or duct to its surroundings is less for refrigerant than for either water or air. As the heat distributed by refrigerant is latent heat, the medium is not as temperature sensitive. The technological advances, particularly in the field of compressors, have led to an increase in the lengths of refrigeration pipe possible in a sytem, as well as the height between units. This in turn has led to the development of split systems using a number of indoor units which can deal with multizone applications.

'Multisplit' system

The simplest system using multiple indoor units is called the 'multisplit' system. This system uses two or more indoor units connected to one outdoor unit.

The refrigeration circuit is similar to that having only one indoor unit. There is a common header or branch to the indoor units, but only one expansion device controlling them all. Although it is a simple application, it has the disadvantage of minimal control over the indoor units which could lead to an imbalance between them.

Multizone modular system

A better system using multiple indoor units is the multizone modular system where one outdoor unit can serve at least eight indoor units each with its own expansion device, leading to better control. As the volume of refrigerant pumped round the system depends on the requirements of the indoor units, this volume varies with the load and hence is also known as a variable refrigerant volume (VRV) system. Such a system can have a simple capillary expansion control with a standard compressor control, but is more usually fitted with an electronic expansion device with inverter control of the compressor. With this latter control, the system is much more responsive to the needs of each zone within a building.

The individual outdoor units can have capacities ranging from 15 kW to 30 kW and indoor units from 2 kW to over 14 kW. The inverter control can change the capacity of the outdoor unit for each compressor in 12–14 steps ranging from 24% to 100%. Special headers or pipe joints ensure that connection between units does not cause problems due to unexpected pressure losses. The indoor units can be cassette, 'built-in', suspended just below the ceiling or mounted at low level, depending on the situation, in a manner similar to single split units. Because of the application of micro-electronic controls to both indoor and outdoor units, the required capacities will be matched and the conditions set in each zone maintained. Like the modern split system, the indoor units can be sited 100 m away from the outdoor unit, with a difference in level of 50 m.

One of the problems with the multizone modular system is that the heat pump is a changeover unit and can only provide either heating or cooling, depending on the mode of the outdoor unit. In many multizone applications simultaneous heating and cooling is required from the plant. This is done in chilled water systems by having more than one heat pump connected to the water pipework, which can then be used to provide all heating or all cooling or split between the two as required. In the closed-loop heat reclaim system it is done by the individual units drawing heat or cooling from the common loop. With refrigeration distribution systems the solution is more complex.

Multizone heat-recovery systems

The need for a number of indoor units, some requiring to heat and others to cool, provides the opportunity to recover heat from those which are cooling and transfer that heat through the refrigeration pipework to those which are heating. Such systems are known as multizone heat-recovery systems (Figure 7.3) and are able to provide simultaneous heating and cooling.

This transfer is done either by two or three pipes from the outdoor unit to special junction boxes where two pipes are taken to each individual indoor unit. Within the refrigeration circuit there are basically three pipes, the hot gas discharge from the compressor which contains the heat from the evaporator but at high temperature, the liquid line from the condenser which contains minimum heat but still at a high temperature and the suction line which also contains the evaporator heat but is at low temperature.

When the system requires all cooling, the discharge line refrigerant is diverted to the outdoor unit coils which are acting as condensers and the suction and liquid lines refrigerant to the indoor units. When the system requires all heating, the discharge line refrigerant is diverted to the indoor units, returning through the liquid line back by the outdoor unit acting as an evaporator. When some of the indoor units are cooling and others heating, the flow to the unit is determined by the special junction boxes sending the refrigerant from the discharge line to those units requiring heat and returning through the liquid line, while those requiring cooling will have refrigerant through the liquid line returning through the suction line. If more units are on heating, then the heat pump will be in the heating mode and part of the outdoor coil will act as an evaporator. If more indoor units are on cooling, the heat pump will be in the cooling mode and part of the outdoor coil will act as a condenser. If the indoor units are in balance, then some will act as condensers and the others as evaporators and the outdoor coil will not be in operation. Usually there is more than one coil and compressor in an outdoor unit, but all will act as part of the same refrigeration system. As all the

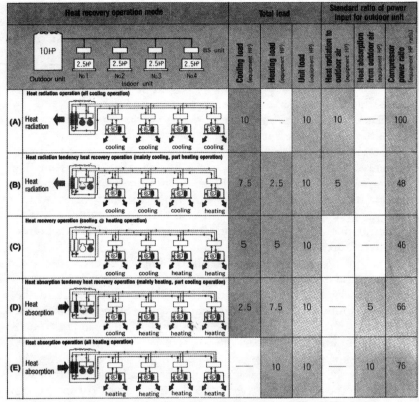

NOTE, Operation modes (A) and (E) are applicable when the outdoor temperature is 35 °C and 0 °C respectively; The other modes are applicable under typical outdoor conditions.

Figure 7.3 *Multizone three-pipe heat-recovery system: (a) heat-radiation mode (all-cooling operation); (b) heat-radiation tendency, heat-recovery operation (mainly cooling, part heating operation); (c) heat-recovery operation (cooling/heating operation); (d) heat-absorption tendency, heat-recovery operation (mainly heating, part cooling operation); (e) heat-absorption operation (all-heating operation) (By courtesy of Daikin Europe NV)*

compressors will be under inverter control, the modulation of control with respect to the load is very good. With three pipes there is a minimum of mixing of the circuits and therefore the best possible economic heat transfer.

The two-pipe multizone heat recovery system works on the principle that the liquid/gas in the liquid line and the discharge gas can mix and then be separated by a special gas/liquid separator. So that in the heating mode, hot gas from the compressor will be delivered to an indoor unit which is heating, the gas is condensed by this unit into a liquid, some of which goes through the expansion device to the gas/liquid separator, while some of the liquid goes to another indoor unit which is cooling to be changed into a gas/liquid to go to the gas/liquid separator. The refrigerant is taken from the separator to the outdoor unit acting in the heating mode, to vaporize the mixture before returning it to the compressor.

When the outdoor unit is in the cooling mode, the discharge from the compressor flows to the outdoor coil to condense into a gas/liquid mixture to go to the gas/liquid separator. There the mixture is separated into a high-pressure gas refrigerant which goes to the indoor unit operating as a heater and a high-pressure liquid refrigerant. The refrigerant from the unit acting as a heater changes into a high-pressure liquid, which then flows through the expansion device of the unit acting as a cooler, is vaporized, and passes through that indoor unit to go to the compressor as a low-temperature low-pressure gas.

The outdoor unit is controlled by an inverter and an outdoor fan control. The mixture of gas and liquid, due to the need for separation, may cause a degree of inefficiency, but the system has the advantage of using only two pipes from the outdoor unit to the indoor units.

A problem with all these refrigerant distribution systems is that the pipework and the indoor units are in the occupied space. There is a need therefore to ensure that in the event of a major leak, where the total charge is discharged into the space it will not exceed the recommended concentrations. In some circumstances individual units may be better than split systems with multiple indoor units. Alternatively, an air- or water-based system may be the best option.

Whatever the system chosen, care must be taken to ensure that the recommendations of the manufacturer with respect to design and installation are followed, as errors in these larger and more complex systems are more difficult and expensive to put right.

EIGHT

Components and controls

For an air conditioning system to be effective it must be capable of operating safely and being controlled within the range of conditions selected. In the case of heat pumps, additional controls are necessary to enable the refrigerant flow to be reversed and the outside air coil to be defrosted when required.

The controls of a unit can be divided into those which are designed to stop the unit operating if a fault occurs or tries to exceed its design range, and those which control the conditions so enabling the system to work. Some controls will be housed within the unit, whereas others may be remote from the unit. There are also certain parts or components of a unit which are not controls but installed to ensure the correct working of the system, such as crankcase heaters, driers, sight glasses and, where required, 'quick couplings'.

Refrigerant components

Crankcase heaters warm the oil to evaporate any liquid refrigerant to avoid dilution of the oil prior to the start of the compressor. When the unit is off and the system equalizes pressures, the refrigerant can condense in the crankcase due to low temperature; liquid refrigerant is not a lubricant which means it can damage compressor bearings. The crankcase heater should always be working for a recommended time before the unit is started and therefore must not be under the control of the starter.

An accumulator can be fitted in the liquid line to prevent liquid refrigerant being carried over into the compressor, known as 'liquid slugging'.

When the refrigerant pipework is installed it is necessary to keep it free of contaminants which includes water. As high-temperature brazing is used during installation oxidization will occur, so to prevent this dry nitrogen is circulated. Dry nitrogen is also used during internal pressure testing. A vacuum must be created in the pipework system to completely dry the system and remove any non-condensables such as air. Driers are installed to absorb any moisture which may enter the system during charging or after installation and to act as filters for other impurities.

The correct refrigerant charge is vital to the proper operation of an air conditioning unit and even more important with a heat pump. The refrigerant charge must be weighed in accurately. Sight glasses are fitted into the liquid line of the system to give an indication of the liquid from the condenser.

In countries where skills are limited, a split-system manufacturer may supply precharged refrigeration pipework in set lengths already charged with the appropriate refrigerant. When precharged refrigerant lines are used in a split system there must be a method of opening the circuit when the unit is connected. This is done by either piercing a disc holding back the refrigerant when screwing the pipes to the parts of the unit, or by Schreader connections. Precharged lines are 'one-shot' couplings and will require refrigeration skills if they are incorrectly fitted and the refrigerant therefore lost.

Safety controls

Safety controls are usually fitted into the unit and consist of high- and low-pressure cutouts, oil pressure switch and compressor start delay.

Pressure-relief devices

With larger systems it is preferable to have pressure-relief devices in the form of a pressure-relief valve or a bursting disc. On smaller units, however, it is considered that pressure-limiting devices are sufficient. This may be subject to change in the future to avoid the risk of refrigerant discharge to the atmosphere.

High-pressure cutouts

A high-pressure cutout is fitted to the high-pressure side of the compressor to ensure that the compressor motor does not operate against excessive pressures. This can take place when the condenser is not removing sufficient heat from the system due, for example, to blockages on the air side, blockages on the refrigerant side or abnormal ambient conditions. The high-pressure cutout is usually manual reset.

Low-pressure cutouts

A low-pressure cutout is located on the low-pressure side of the compressor and is intended to ensure that the evaporator pressure, and hence its temperature, does not fall too low and thereby cause icing on the air side of the coil, and to prevent vacuum conditions which are detrimental to the hermetic motor. The low-pressure cutout is normally auto reset. Often both the high- and low-pressure cutouts are combined into one control.

Other safety controls

A de-ice thermostat or sensor may also be fitted to the evaporator coil to prevent ice formation on the air side.

An oil pressure switch may be fitted to stop the compressor operating should the oil pressure drop to below a predetermined point.

The compressor start delay is designed to ensure that the compressor will not restart until the refrigerant pressures have equalized throughout the system. A typical setting is 3 minutes.

Operational controls

Operational controls can be split between those which deal with the refrigeration circuit, those dealing with other parts of the unit and finally those which control the condition of the air supplied to the space.

The controls dealing with the refrigeration circuit are the expansion valve or device and possibly a head pressure controller. In the case of heat pumps, additional controls are incorporated such as check or non-return valves, reversing valves and defrost controls.

Expansion valves
Expansion valves meter the flow of refrigerant and help to maintain evaporation within set limits and prevent the evaporator discharging liquid along with the vapour to the compressor, which could cause damage to the compressor. The thermostatically controlled expansion valve has a bellows-operated needle valve, operated by a secondary refrigerant-filled sensor connected to the bellows by means of a capillary tube. The sensor is attached to the evaporator outlet so that should the evaporator outlet gas temperature rise the valve will open, increasing the refrigerant flow, or the reverse when the temperature falls. Because the valve can control only the evaporator to which the sensor is fitted for a reverse-cycle heat pump, where either coil can be an evaporator, an additional expansion valve is needed on the other coil. The expansion valve on air conditioning applications will also have a pressure-equalizing line to compensate for evaporator pressure drop.

In the case of a heat pump where there are two expansion valves, controls are needed to ensure that the one situated on the coil, which is acting as a condenser, is bypassed. This is done by a check valve fitted across the expansion valve, allowing the refrigerant to bypass in one direction but be diverted through the expansion valve in the other.

Reversing valve
To reverse the cycle it is necessary to direct the flow of refrigerant from the compressor to one coil or the other, depending on the mode of operation. This is done by means of a solenoid-operated four-port reversing valve which is usually a servo-operated piston type to give a smooth changeover.

Defrost control

When an air-source heat pump operates in the heating mode during the winter, the outside coil will ice up due to moisture in the ambient air. In time the ice build-up will be such that the flow of air across the coil is severely restricted. At such times it is necessary to defrost the coil. This can be done by bypassing some of the hot gas from the compressor to the coil, reversing the cycle or by an electric heater on the coil. Defrost is initiated by a defrost control which can either sense the increase in pressure across the coil caused by the ice build-up and/or time/temperature requiring the control to operate at preset intervals should the thermostat contacts be closed.

Head pressure control

The head pressure control is a valve which responds to a falling pressure and is fitted at the outlet of the condenser coil. Falling pressure can occur when the outside temperature drops and this control helps to maintain capacity under such circumstances.

Low ambient control

Of the controls that deal with the other parts of the unit, the low ambient control is one of the most important for systems installed in the UK. Where the unit is to operate in the cooling mode during the cooler weather, normally below 10°C, some form of air flow control needs to be fitted to the condenser air fan. Due to the lower ambient temperature, which will be less than that for which the unit was designed, the condenser coil will have a greater temperature difference. Because of this the coil will have a higher heat transfer rate and so will try to remove more heat from the refrigerant. This in turn will affect the balance of the refrigeration cycle and force it to reduce pressures and temperatures outside its range. To avoid this, the amount of air passing over the condenser needs to be reduced to restore the balance. This is achieved by intermittent operation, where there is only one fan, or operating only one of the fans where there is more than one, or by some form of speed control. These will be initiated either by a pressure switch in the condenser outlet or by a thermostat sensing liquid refrigerant temperature.

Thermostats, humidistats and fan-speed controllers

The controls dealing with the condition of the supply air to the space are thermostats, humidistats and fan-speed controllers.

The thermostats used for air conditioning and heat pumps will actually be two thermostats within the one case: one opening on a rising temperature controlling heating and the other closing on a rising temperature controlling cooling. There is often an adjustable differential between one opening and the other closing to avoid the unit 'hunting' between the two modes of operation.

The thermostat can either be integral within the unit, usually sensing the return air from the space, or remote and located within the conditioned space.

Where humidity control of a space is important, then a humidistat is used to control the compressor. This is a switch similar to a thermostat but which responds to changes in humidity. When it is used to control the unit the air supplied is usually too low for comfort; therefore a thermostat controlling a heater is incorporated into the system to reheat the air to an acceptable temperature.

Both temperature and humidity controls can only control the condition at the sensor and not over all the conditioned space. The sensitivity of the control will also only be that at the sensor, whereas the actual control of the space will be less sensitive and may be subject to greater fluctuations.

Often the capacity needs a boost at start-up and the noise levels reduced at other times. This is achieved by some form of fan-speed control, which controls at high speed to boost the capacity, medium speed for normal operation and low speed when quiet operation is preferred.

A programmed time switch can also be included to ensure the system is only operated when required, thus reducing energy costs.

Capacity control

One of the problems of direct expansion systems is how to change the output from the unit to match a changing load. Where the unit has multiple compressors, this can be achieved by 'offloading' the compressors. Normally, however, the unit operates on an on/off basis. Attempts have been made to vary the speed of the compressor and hence its output by voltage control, but this has not proved satisfactory. With the advances in electronics, another form of capacity control has been developed known as inverter control. The normal 50-cycle a.c. electricity supply is changed into a d.c. supply, this in turn is changed to a variable frequency a.c. supply. By varying the frequency, the speed of the motor varies from about 20% to 130% giving a similar variation in capacity. Although more expensive than normal control, the savings in energy consumption will pay the extra cost and because the packaged unit can be 'overrun' for short periods, less units need to be installed to deal with a given load.

Single electronic controller

Many of the functions already discussed can be incorporated into one single electronic controller, and more and more packaged units are being fitted with such devices as standard.

These controllers now incorporate additional features such as matching compressor speed controls with electronic expansion valves, discharge air temperature sensors to smooth air temperature, sleep controls to reduce fan speeds and hence noise, and stepped timer functions.

Not only can they deal with the operational control of a number of units, but can also report on faults within the system and the status of the system. Because of the microchip, these controllers will also incorporate memories, allowing for the system to be programmed and the fault history to be recorded.

Future controllers

The controls of air conditioning plant will in the future become more and more sophisticated owing to the application of microelectronics and will be made to control units in such a way that packaged equipment will be capable of modulating to the same range as larger central plant. Future controllers will also ensure that discharges of refrigerant into the atmosphere will be contained to the required levels or that suitable warning is given.

NINE

Distribution and air movement

The effects of introducing cool air

With the advances in packaged air conditioning equipment, the
need for a ducted system in a building has been reduced. At one
time the choice was for either window units, free-standing units or
ducted units. If good distribution of the conditioned air was
required, then a ducted system was the only way to achieve this,
particularly if high-level diffusion was necessary.

Now, however, there is a range of units – such as the console,
high-level mounted, ceiling mounted or cassette – which enable
the designer to satisfy most requirements. But even these modern
units require, on the part of the designer, an understanding of
distribution and air movement to avoid an unsatisfactory application.

The idea of an air conditioning system is to provide a supply of
air into a space which will deal with the environmental load
requirements of the space and at the same time remove exhausted
air for air treatment. If cooling and dehumidifying is required,
then the supply air should be cooler and drier than the space
conditions. If heating is required, the supply air will be warmer
than the space conditions. The amount of air and its temperature
will depend on many factors, but it should be introduced into the
space in such a way that discomfort is avoided.

94

Also, the treated air must be supplied in proportion to the section of the space where the load is generated. Frequently this may dictate the numbers, type and size of units within any given space or room. Obviously the cheapest solution is likely to be one large unit, but three may be needed to meet the space comfort requirements.

Any space will have natural air currents caused, for example, by cold down-draughts from windows during the winter or warm rising air currents from heat-producing machines. These currents help to promote convection from the body which can produce a feeling of freshness within the room without producing discomfort.

Air introduced by an air conditioning system or unit should not unduly interfere with these natural air currents within the occupied zone, except to overcome a particular problem, such as down-draughts. To avoid this, it is necessary to consider the location of the supply air inlet, and the direction, velocity and temperature of the supply air. Although the location of outlets for exhaust air or return air is important, they have less effect over natural air currents in low-pollution areas. In contaminated areas, such as where heavy smoking occurs, separate extraction may be required and the return air location is important.

To ensure that the effects of the conditioned air are felt all over the space, it is essential to have an inlet device which will spread the air evenly where required. Such a device is a diffuser and can either be located in the distribution ducting or as an integral part of the room unit.

Understanding the characteristics of air

There are certain characteristics of air which must be considered when selecting the location of inlet diffusers. Hot air rises and cold air drops, due to the difference in densities between the air in the space and that of the supply air. As an air conditioning system will be providing both heating and cooling, there is an immediate conflict of densities, and the choice has to be a compromise.

However, the choice can be assisted by selecting the right direction and velocity of the airstream leaving the diffuser. These two characteristics will, to an extent, overcome the effect of density by forcing the air in the required direction. As the volume of air needed to deal with a given load is normally greater for cooling than heating, diffusers are usually designed for this larger volume. The effect of these larger volumes when heating is to reduce the leaving temperature and hence the effect of the differing densities of air.

Diffusion characteristics

Where possible, mixing of the supply air should occur outside the occupied zone, i.e. the space which extends from the floor to a height of about 2 m.

Room air should be entrained by the air coming from the inlet diffuser to ensure proper mixing. The air leaving the diffuser is known as the primary air and entrained room air is known as the secondary air (Figure 9.1).

The primary air has a velocity given to it by the fan in the unit or the system and is normally not less than 0.75 m/s. This velocity imparts an energy to the airstream, enabling it to overcome the resistance of the surrounding room air, to entrain the room air and enable the mixed air to reach the distant parts of the space. As the airstream travels across the space its velocity decreases and the temperature becomes closer to room temperature. The distance travelled from the diffuser by the airstream until its velocity is reduced to about 0.25 m/s, known as the terminal velocity, is called the 'throw'. The lower the initial velocity, the shorter the throw; the greater the velocity, the longer the throw. If the throw is too great, then there is a tendency for the air to hit walls, etc., and create draughts. In ducted systems a diffuser can be selected with the right characteristics, but the preset throw of a diffuser incorporated in a unit may limit its application and location.

As the velocity of the air from the system decreases, it entrains more room air and two other characteristics occur. First, when

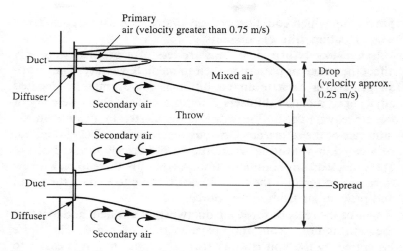

Figure 9.1 *Air flow pattern from high-level diffuser*

cooling, the air tends to drop and secondly, the airstream becomes wider than that leaving the diffuser. The airstream should not drop into the occupied zone before reaching its terminal velocity, and this will therefore limit mounting height of the diffuser and location to avoid ceiling obstructions. Also, one airstream should not spread so that it affects an adjacent airstream from another unit or diffuser. These characteristics will limit either the throw or the location of the diffuser. In addition, the spread of an airstream affects the throw, in that to obtain a wider spread the diffuser vanes are arranged to spread the air pattern, thereby increasing the resistance to flow, which in turn reduces the volume and initial velocity.

The effect of any airstream hitting another is to change both to flow in a different direction. When the airstream is colder than room air, that direction is usually downwards, creating a draught. Also, the problem that could be created by beams across the ceiling, light fittings, hanging promotional signs or any other protrusion which could deflect the airstream must be considered when locating diffusers or units.

Exhaust or return air outlets are less of a problem and only affect the room air pattern in their immediate vicinity. The return air grille in a cassette air conditioner, for example, is located adjacent to the inlet diffusers with no problem, as the velocity of the air leaving the diffusers directs the airstream away from the influence of the return air. Obviously, return air grilles should not be located in a position where they can be easily blocked or are able to pick up any contamination. As they are there to pick up air to return to the unit they should not be able to bypass the room and pick up air from other sources.

One of the reasons for using ductwork in a system and locating the unit remote from the space is that of noise. The noise generated by the unit is away from the space and it is easier to provide attenuation if there is any remaining in a ducted system. However, selecting the wrong diffuser or grille, leading to an unacceptable velocity, will also create noise.

The diffusion characteristics of units

Each type of packaged unit has its own air movement characteristics and these should be taken into account when considering the overall distribution requirements of the system.

The window unit is simple, but has the poorest distribution. Usually the supply air is discharged from fixed louvres horizontally from the front of the unit. If there is any adjustment that can be made to the diffuser, then it will be very limited. When installed on the window sill at low level, it blows cold air across the room in the occupied zone. When installed at high level, these problems are minimized.

The air flow characteristics are similar to Figure 9.1, but because the diffusers are basic there is less control.

The console unit has better distribution in that the air flow is vertical from the top of the unit, directing the air away from the occupied zone. It can also be directed into an area of high heat gain if placed below a window. The air flow should follow a line up the wall, across the ceiling and then drop into the occupied zone. One problem could be that, in an office, documents may inadvertently be put over the outlet, blocking the air path.

The free-standing unit usually discharges from the top horizontally over the occupied zone and can provide satisfactory distribution for units up to 25 kW. Units above 25 kW should only be considered if the background noise is high, such as in industrial applications, or alternatively used with ducting. With larger sizes, because of the great volumes of air flowing through one diffuser it is difficult to avoid dropping cold air into the occupied zone before it reaches its terminal velocity. The free-standing unit should be mounted on a floor plenum to increase the height of discharge and care is needed to ensure that occupants are not seated near to the common return air point.

Normally, roof units or ceiling void units will be ducted, although some manufacturers have a special diffuser box and roof-mounting kits which avoids the use of ducting when used in shopping centres, for example, with smaller roof units.

Low-level split units will have vertical distribution similar to the console unit. High-level split units will have horizontal discharge below the ceiling above the occupied zone. The air pattern will depend on how near the discharge is to the ceiling. If it is close, it will tend to flow across at high level for longer distances than when sited away from the ceiling. Some high-level units have the air discharge below the return air and are only suitable for the higher ceiling heights.

Ceiling-mounted units provide a horizontal discharge at ceiling level. This makes the best use of the coanda effect (see Appendix I), keeping the airstream out of the occupied zone until it is properly mixed with the room air.

Cassette units are mounted at high level from the ceiling or within the ceiling. Where the discharge is immediately below a ceiling, the air flow maintains its horizontal direction and delays dropping into the occupied zone, compared with the discharge from the same unit when mounted away from the ceiling at high level.

Obviously there is no point in trying to cool the volume above the occupied space where there is a high ceiling or roof space, and there is a need to avoid hot air rising before it reaches the occupied space. Therefore, in these situations, to avoid cooling the volume above the occupied zone the discharge should be at high level to the occupied zone. It should be mounted at minimum recommended height, taking into account the drop characteristics of the discharge. This will also help to reduce the throw needed by the warm air to reach the occupied space.

Although most packaged air conditioning requirements can be resolved using individual units or split systems having indoor room units, there are occasions when the design can only be satisfied using a ducted distribution system. It will then be necessary to layout and to size such a distribution system. This will enable the duty of the air conditioning unit fan to be specified or to check that the standard fan supplied is of sufficient size to cope with the volume and resistance of the system. The air diffuser requirements, noise problems and fan characteristics are all similar for ducted systems as for those of individual units already discussed.

Duct sizing

Duct sizing can be carried out in a number of ways, but that most used for smaller packaged air conditioning systems is known as the 'equal friction' method. This method is sufficiently accurate for all applications using small conventional low-velocity ductwork. With this method, a nominal resistance per unit length is chosen suitable for the particular application and each section of ductwork is sized, as far as possible, to that resistance. In practice, because the variation in duct sizes is limited, one section may have a resistance per unit length above that chosen, whereas another may have less. The idea is for the total pressure drop of all sections to average the resistance chosen. The more equal the section pressure drops, the nearer the system will be proportionately balanced, otherwise secondary means of balancing such as dampers may need to be introduced. The type of ducting, its construction and installation should conform with local codes, and a minimum guide is typically the Heating and Ventilating Contractors' Association series of 'Guides to Good Practice'.

Generally, the lower the resistance per unit length, the larger the duct size will be for a given volume, but as the velocity will be less the quieter the air flow will be through it. Therefore, a low resistance should be picked for residential applications or conference rooms, a higher resistance for shops and offices and a still higher resistance for industrial applications.

The resistance or pressure drop is measured in pascals per metre (Pa/m) and the following are the typical recommended values for simple systems relating to general applications:

(a) Quiet   0.4 Pa/m
(b) Commercial   0.6 Pa/m
(c) Industrial   0.8 Pa/m

Before sizing the ductwork, its layout must be determined and this layout will be a function of the size, type and number of diffusers needed to distribute the conditioned air into the space. This in turn is determined by the throw, spread and drop characteristics of the diffusers chosen when handling different air volumes. Having selected and positioned the required diffusers, they will be connected by sections of ductwork back to the unit. The return air to the unit will also be ducted to the unit from a suitable location. The index circuit of the system is determined and it is only the resistance of this circuit which is used to determine the fan pressure. The index circuit is the one which will have the greatest resistance to air flow and is normally the longest run of ductwork. The circuit is from the room or space, through the unit, out through the supply air ducting to the furthest diffuser and then back into the room. Sometimes, because of extra resistance caused by fittings, such as bends, the index circuit may not be the longest run.

The selected pressure drop will give the resistance to air flow of the straight lengths of duct, and the manufacturer's data on the diffuser, related to the air flow, will give the resistance through the

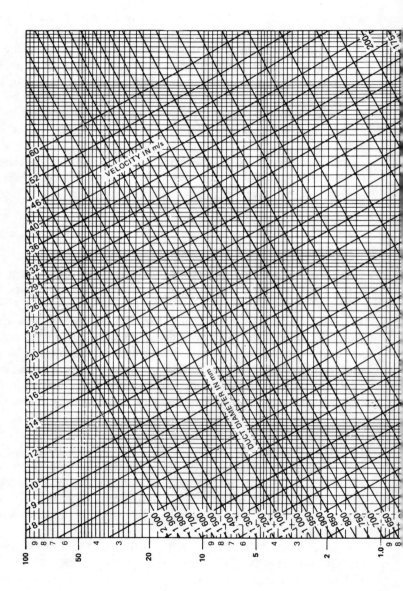

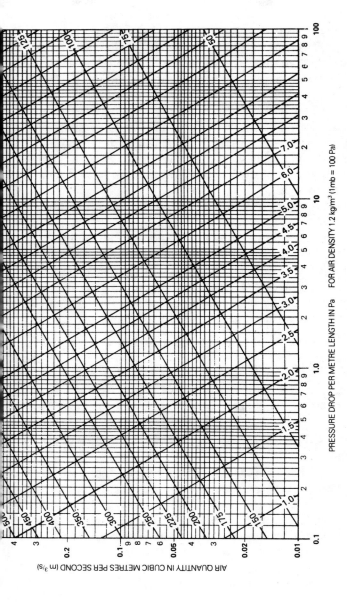

Figure 9.2 *Duct-sizing chart (By courtesy of the Chartered Institution of Building Services Engineers)*

diffuser. Similar data will give the resistance of the return air grille. There is also the resistance of the fittings and changes in duct size to be considered. Whenever there is a change in direction, or an increase or decrease in duct size, causing a change in velocity or the entrance or exit from ducts, there will be resistance to the flow of air.

The duct size is determined by using the duct-sizing chart in Figure 9.2. This chart gives a series of pressure drops per metre on the horizontal axis, air volumes (in m³/s) on the vertical axis, duct diameters (in mm) on the lines sloping down from right to left and velocities in the section of duct selected (in m/s) on lines sloping down from left to right.

The circular duct sizes selected can be changed into equivalent rectangular duct sizes by using Table 9.1.

The resistance to air flow through fittings is found by multiplying the velocity pressure of the fitting by a '*K*' factor related to that particular fitting. Table 9.2 gives the conversion of the velocity of a duct section to a velocity pressure, while Figure 9.3 gives the '*K*' factors for most normally used fittings.

To size a duct layout use the following procedure:

1 Select the diffusers to be used. Note their resistance at the system conditions.
2 Sketch the layout from the unit to each diffuser.
3 On the sketch, mark out the duct sections, lengths of ducts and the various fittings. Label each section so that it can be identified.
4 Show against each terminal section or branch the air volume from the diffuser.
5 Add the volumes back to the unit and from the unit through the return air and any fresh air inlet. Note the volume in each duct section.
6 Having selected a pressure drop suitable for the application, select a duct diameter appropriate to the air volume of the particular duct section. Note this diameter and the equivalent velocity. Multiply the duct section length by the resistance or pressure drop to find the resistance for that section.

7 If rectangular ducts are to be used, then apply Table 9.1 to arrive at a size. Square section ducts provide the least resistance, so make the selection as near to square as possible. Avoid widths of ducts greater than twice the height of the duct. Also, when changing from one section to another only change one side of the duct to avoid expensive duct formations.

8 Add up each section resistance to arrive at the total resistance of the straight ducts.

9 Note the types of fittings and select suitable '*K*' factors from Figure 9.3, and multiply these by the velocity pressure determined from the velocity of the air through the fitting. This will give the resistance of each fitting or change section. In the case of bends or tee-pieces it may be necessary to work out the ratios needed to determine the actual '*K*' factor. However, in most small applications the simplified values should be sufficient.

10 Add up the values of the resistances to give the total resistance through all fittings.

11 Add together the resistance given by the manufacturer of the diffuser and return air grille, with that of the straight ducts and the fittings, to give the total resistance of the duct layout. Only the one supply diffuser on the index circuit should be included.

12 The external static pressure of a unit capable of overcoming the resistance in the ductwork system is based on the unit filter being clean. This is rarely the case in normal operation, so additional resistance should be added to that calculated figure. This should be about 50 Pa for units up to 25 kW capacity and 120 Pa for units above.

13 The total air volume is used to determine the fan duty, i.e. the volume from all outlets. However, only the resistance through the index circuit will be used to specify the required external static pressure.

Some problems with diffusers and ducting

If the temperature of the air surrounding the duct is higher than that in the duct, cooling of the air surrounding the duct can occur.

105

Table 9.1 Equivalent rectangular duct sizes (By courtesy of the Chartered Institution of Building Services Engineers)

Dimen. of side, b	Dimension of side of duct, a																			
	100	125	150	175	200	225	250	300	350	400	450	500	550	600	650	700	750	800	850	900
100	110	123	134	145	154	163	171	185	199	211	222	232	242	251	260	268	276	284	291	298
125	347	138	151	162	173	183	192	209	225	239	251	263	275	285	295	305	314	323	331	339
150	385	394	165	178	190	202	212	231	248	264	278	291	304	316	327	338	348	358	368	377
175	421	430	440	193	206	218	230	251	269	287	303	317	331	344	357	369	380	391	401	411
200	454	464	474	484	220	233	246	269	289	308	325	341	356	371	384	397	409	421	433	444
225	485	496	507	517	527	248	261	286	308	328	346	364	380	395	410	424	437	450	462	474
250	515	527	538	549	560	570	275	301	325	346	366	385	402	419	434	449	463	477	490	503
300	570	583	596	608	620	632	643	330	357	381	403	424	443	462	479	496	512	527	542	556
350	620	635	649	662	676	689	701	714	385	412	436	459	481	501	520	539	556	573	589	605
400	667	683	698	713	727	741	755	768	794	441	467	492	515	537	558	578	597	616	633	650
450	710	727	744	760	776	791	806	820	848	874	496	522	547	571	594	615	636	655	674	693
500	751	770	787	804	821	837	853	869	898	927	954	551	577	603	627	650	672	693	713	733
550	790	810	828	847	864	882	898	915	946	976	1005	1033	606	633	658	682	706	728	749	770
600	827	848	867	887	905	924	941	959	992	1024	1054	1084	1112	661	688	713	738	761	784	806
650	862	884	905	925	945	964	982	1001	1036	1069	1101	1132	1162	1191	716	743	769	793	817	840
700	896	919	940	962	982	1002	1022	1041	1077	1113	1146	1179	1210	1240	1269	771	798	824	849	873
750	928	952	975	997	1018	1039	1059	1079	1118	1154	1190	1223	1256	1287	1318	1347	826	853	879	904
800	959	984	1008	1031	1053	1075	1096	1117	1157	1195	1231	1267	1301	1333	1365	1396	1426	881	908	934
850	989	1015	1039	1063	1086	1109	1131	1152	1194	1234	1272	1308	1344	1378	1411	1443	1474	1504	936	963
900	1018	1044	1070	1095	1119	1142	1165	1187	1230	1271	1311	1349	1385	1421	1455	1488	1520	1551	1582	991
950	1046	1073	1100	1125	1150	1174	1198	1221	1265	1308	1349	1388	1426	1462	1498	1532	1565	1597	1629	1659
1000	1101	1128	1155	1180	1205	1230	1254	1299	1343	1385	1426	1465	1503	1539	1575	1609	1642	1675	1706	
1050	1156	1184	1210	1236	1261	1285	1332	1378	1421	1463	1503	1542	1580	1616	1652	1686	1719	1752		
1100	1211	1239	1265	1291	1316	1365	1411	1456	1499	1540	1580	1619	1657	1693	1729	1763	1797			
1150	1266	1294	1320	1346	1396	1444	1490	1534	1577	1618	1658	1696	1734	1770	1806	1840				

$$d = 1.265 \left[\frac{(ab)^3}{a+b}\right]^{0.2}$$

Dimen. of side, b	\	\	\	\	\	1200	1250	1300	1400	1500	1600	1700	1800	1900	2000	2100	2200	2300	2400	2500
1200						1322	1349	1375	1427	1476	1523	1568	1612	1654	1695	1735	1773	1811	1847	1882
1250							1377	1404	1456	1507	1555	1602	1646	1690	1732	1772	1812	1850	1888	1924
1300								1432	1486	1537	1587	1634	1680	1725	1768	1809	1850	1889	1927	1965
1400									1542	1596	1648	1697	1745	1792	1837	1881	1923	1964	2004	2043
1500										1652	1706	1758	1808	1857	1904	1949	1993	2036	2078	2119
1600											1762	1816	1868	1919	1968	2015	2061	2106	2149	2192
1700												1872	1926	1979	2029	2078	2126	2173	2218	2262
1800													1982	2036	2089	2140	2189	2237	2284	2330
1900														2092	2147	2199	2250	2300	2348	2396
2000															2203	2257	2310	2361	2411	2459
2100																2313	2367	2420	2471	2521
2200																	2423	2477	2530	2582
2300																		2533	2587	2640
2400																			2643	2697
2500																				2753
	950	1000	1050	1100	1150	1200	1250	1300	1400	1500	1600	1700	1800	1900	2000	2100	2200	2300	2400	2500

Dimension of side of duct, a

Table 9.2 Conversion of velocities to velocity pressures (By courtesy of the Chartered Institution of Building Services Engineers)

Velocity m/s	0·0	0·1	0·2	0·3	0·4	0·5	0·6	0·7	0·8	0·9
0	0·00	0·01	0·02	0·05	0·10	0·15	0·22	0·29	0·38	0·49
1	0·60	0·73	0·86	1·01	1·18	1·35	1·54	1·73	1·94	2·17
2	2·40	2·65	2·90	3·17	3·46	3·75	4·06	4·37	4·70	5·05
3	5·40	5·77	6·14	6·53	6·94	7·35	7·78	8·21	8·66	9·13
4	9·60	10·09	10·58	11·09	11·62	12·15	12·70	13·25	13·82	14·41
5	15·00	15·61	16·22	16·85	17·50	18·15	18·82	19·49	20·18	20·89
6	21·60	22·33	23·06	23·81	24·58	25·35	26·14	26·93	27·74	28·57
7	29·40	30·25	31·10	31·97	32·86	33·75	34·66	35·57	36·50	37·45
8	38·40	39·37	40·34	41·33	42·34	43·35	44·38	45·41	46·46	47·53
9	48·60	49·69	50·78	51·89	53·02	54·15	55·30	56·45	57·62	58·81
10	60·00	61·21	62·42	63·65	64·90	66·15	67·42	68·69	69·98	71·29
11	72·60	73·93	75·26	76·61	77·98	79·35	80·74	82·13	83·54	84·97
12	86·40	87·85	89·30	90·77	92·26	93·75	95·26	96·77	98·30	99·85
13	101·40	102·97	104·54	106·13	107·74	109·35	110·98	112·61	114·26	115·93
14	117·60	119·29	120·98	122·69	124·42	126·15	127·90	129·65	131·42	133·21
15	135·00	136·81	138·62	140·45	142·30	144·15	146·02	147·89	149·78	151·69
16	153·60	155·53	157·46	159·41	161·38	163·35	165·34	167·33	169·34	171·37
17	173·40	175·45	177·50	179·57	181·66	183·75	185·86	187·97	190·10	192·25
18	194·40	196·57	198·74	200·93	203·14	205·35	207·58	209·81	212·06	214·33
19	216·60	218·89	221·18	223·49	225·82	228·15	230·50	232·85	235·22	237·61
20	240·00	242·41	244·82	247·25	249·70	252·15	254·62	257·09	259·58	262·09
21	264·60	267·13	269·66	272·21	274·78	277·35	279·94	282·53	285·14	287·77
22	290·40	293·05	295·70	298·37	301·06	303·75	306·46	309·17	311·90	314·65
23	317·40	320·17	322·94	325·73	328·54	331·35	334·18	337·01	339·86	342·73
24	345·60	348·49	351·38	354·29	357·22	360·15	363·10	366·05	369·02	372·01
25	375·00	378·01	381·02	384·05	387·10	390·15	393·22	396·29	399·38	402·49
26	405·60	408·73	411·86	415·01	418·18	421·35	424·54	427·73	430·94	434·17
27	437·40	440·65	443·90	447·17	450·46	453·75	457·06	460·37	463·70	467·05
28	470·40	473·77	477·14	480·53	483·94	487·35	490·78	494·21	497·66	501·13
29	504·60	508·09	511·58	515·09	518·62	522·15	525·70	529·25	532·82	536·41
30	540·00	543·61	547·22	550·85	554·50	558·15	561·82	565·49	569·18	572·89
31	576·60	580·33	584·06	587·81	591·58	595·35	599·14	602·93	606·74	610·57
32	614·40	618·25	622·10	625·97	629·86	633·75	637·66	641·57	645·50	649·45
33	653·40	657·37	661·34	665·33	669·34	673·35	677·38	681·41	685·46	689·53
34	693·60	697·69	701·78	705·89	710·02	714·15	718·30	722·45	726·62	730·81
35	735·00	739·21	743·42	747·65	751·90	756·15	760·42	764·69	768·98	773·29
36	777·60	781·93	786·26	790·61	794·98	799·35	803·74	808·13	812·54	816·97
37	821·40	825·85	830·30	834·77	839·26	843·75	848·26	852·77	857·30	861·85
38	866·40	870·97	875·54	880·13	884·74	889·35	893·98	898·61	903·26	907·93
39	912·60	917·29	921·98	926·69	931·42	936·15	940·90	945·65	950·42	955·21
40	960·00	964·81	969·62	974·45	979·30	984·15	989·02	993·89	998·78	1003·69

Radius Bends — Factors refer to V.P. in the duct							Mitre Bends — Factors refer to V.P. in the duct	

Rectangular Ducts θ = 90°

$\frac{h}{w}$	Radius Ratio r/w				
	0.75	1.0	1.5	2.0	3.0
0.2	0.64	0.47	0.44	0.40	0.43
0.4	0.55	0.36	0.33	0.28	0.31
0.5	0.45	0.29	0.22	0.20	0.22
0.7	0.41	0.26	0.18	0.16	0.18
1.0	0.39	0.23	0.14	0.13	0.14
1.5	0.41	0.21	0.12	0.11	0.11
2.5	0.45	0.20	0.11	0.09	0.10
4.0	0.51	0.22	0.11	0.10	0.10
6.0	0.62	0.26	0.14	0.11	0.11

Mitre bends: 1.25, 1.22, 0.72, 0.67, 0.35, 0.71

Round Ducts θ = 90°

No. of pieces	Radius Ratio r/d				
	0.75	1.0	1.5	2.0	3.0
3	0.58	0.46	0.40	0.42	0.46
4	0.56	0.42	0.34	0.32	0.34
5	0.50	0.36	0.30	0.26	0.26
radius	0.45	0.34	0.24	0.23	0.22

Bends in Close Series (Factor for multiple arrangement as percentage of equivalent single bend)

150% of one bend, 200% of one bend, 240% of one bend

Duct Entries — Factors refer to V.P. in the duct: 0.43

Gradual	θ	k
	30°	0.02
	45°	0.04
	60°	0.07

45° Louvres F.A.	k_j	k_b
0.9	—	—
0.8	1.40	0.40
0.7	2.10	0.80
0.6	3.00	1.30
0.5	4.50	2.30

Open Dampers	k
	0.5
	0.2

Figure 9.3 *Velocity pressure 'K' factors for most fittings (By courtesy of the Chartered Institution of Building Services Engineers)*

Branch Pieces Factors refer to VP in the offtake					
Round Ducts (Flow to branch)	$\frac{V_3}{V_1}$	Area Ratio A_3/A_1.			
		= 90°		= 45°	
		0.15	0.50	0.15	0.50
	0.6	3.70	1.10	2.40	0.66
	0.8	2.30	0.49	1.50	0.32
	1.0	1.70	0.34	1.10	0.21
	1.1	1.50	0.32	0.95	0.20
	1.2	1.40	0.30	0.88	0.19
	1.3	1.30	0.32	0.80	0.20

Rectangular Ducts (Flow to branch)	$\frac{V_3}{V_1}$	Area Ratio A_3/A_1		
		0.15	0.50	.90
	0.6	1.10	—	—
	0.8	0.65	0.25	0.52
	1.0	0.40	0.20	0.25
	1.1	0.40	0.18	0.25
	1.2	0.35	0.18	0.33
	1.3	0.36	0.19	—

GENERAL NOTES

BEND ANGLES Where bends turn through angles of less than 90°, the factor k may be presumed to vary in proportion to θ/90°.

CHANGES OF SHAPE (transitions) For tapered changes of shape where θ <60° and $A_1 ≙ A_2$, the factor may be taken as 0·15.

SPLITTERS Where straight ducts have splitters, the straight duct friction loss through each component part should be considered.

APPROACHES The values for the factor k, here quoted, assume that the approaching velocity profile is regular. Any eccentricity or distortion may increase or decrease the loss, as illustrated for bends in series.

Figure 9.3 *cont.*

Under certain conditions, condensation on the outside of the duct walls will take place. The psychrometric chart can be used to check if this will happen and therefore if there is a need to insulate the duct, including a vapour barrier to stop the moisture permeating the insulation.

If a diffuser is placed in the wall of the duct without a throat or spigot, there is a possibility that air could be drawn in by the diffuser as well as being discharged. The air passing upstream of the diffuser creates a low-pressure zone, so that about a third of the diffuser area becomes an inlet to the duct. This reduces the effectiveness of the diffuser as well as altering the characteristics such as spread and drop. Deflecting or turning vanes in the duct, directing part of the airstream through the diffuser, will minimize this tendency if it is impossible to provide a spigot.

Apart from excessive air noise, which can be overcome by the correct selection of pressure drop and diffuser characteristics, vibration from the unit can cause noise in the system. To avoid the transmission of vibration, some form of isolation such as flexible connections can be made between the supply and return air ducts and the unit. Also the unit may require anti-vibration mountings or hangers. Further reductions in noise level can be achieved by acoustic treatment such as attenuators, acoustic louvres or lining the ducts with fire-resistant acoustic material.

It is important that the system conforms to any building fire requirements or other regulations. For example, where ducts pass through fire walls, fire dampers with suitable access doors will be required. Also, transfer grilles to corridor may require fire dampers.

TEN

Proposals

Information required to select the appropriate unit

Before putting forward to the client the design considered to be correct for the application, information is needed about space, its use and the client's requirements.

Obtaining the correct information is vital to a proper assessment of the problem and to the determination of the best solutions. Time spent in asking the right questions and conducting an adequate survey can ensure that the design will meet the needs of the occupier and is both economic and energy efficient.

The first question to be asked should be: 'Is air conditioning really the solution to the problem or can it be solved by other means?' For example, if the problem is too much moisture and not overheating, then a dehumidifier may be the more effective solution. However, if this need is combined with overheating, then air conditioning will be the answer. Although mechanical ventilation can provide a degree of relief from overheating, it cannot easily deal with the problems of high humidity.

Also consider what can be done to the space to reduce the need for air conditioning and hence reduce the size of the equipment required. One example could be to use awnings or solar reflecting film to reduce the solar gain. Another is to consider precooling and allowing the temperature to float outside the accepted range for short periods. These matters have been discussed in other chapters.

The following is a shortlist giving the major areas where information is needed to enable a proper assessment to be made:

A The building.
1 Dimensions.
2 Structure: windows, walls, floor, ceiling.
3 Orientation.
4 Shading: by awnings, blinds or other buildings.
5 Location: residential, commercial, town or suburb.

B Occupancy and internal loads.
1 Numbers of people: average or maximum.
2 Type of work: light, heavy, sitting, standing or walking.
3 Internal noise levels.
4 Layout of equipment or furniture.
5 Lighting: type and wattage.
6 Equipment: type and wattage.
7 Use of equipment and lighting.

C Services.
1 Electrical power.
2 Drainage.
3 Water if required.

D Equipment.
1 Possible locations.
2 Limitations affecting type.
3 Maintenance problems.
4 Acceptable external noise levels.
5 Obstructions to air flow.

E Conditions.
1 Set conditions for temperature and, if required, humidity.
2 Additional ventilation requirements.
3 Special filtration.
4 Hours of operation.

F Special requirements.
1 Planning regulations.
2 Environmental considerations relating to noise.
3 Permission to use water or drainage.
4 Adequacy of the electricity supply.

Estimation of cooling load and consumption

Having obtained the above information and used it to design a suitable system, it is then necessary to put forward these proposals to the client.

They should be set out in such a way that the client understands what is to be provided and what is required of the client, as too often there is misunderstanding in this area which can lead to dissatisfaction with the system offered, even though it is satisfactory.

To avoid these misunderstandings the following minimum information should be provided with the proposals.

1 Design
The temperature conditions both inside and out upon which the design is based. If 'comfort cooling' only is proposed, an idea of the probable temperature difference between inside and out. This will give the customer an idea of the type of conditions to expect – very often the customer believes that the air conditioning plant will provide cool conditions regardless of the external temperatures. It also ensures that at a later date these facts can be checked should dissatisfaction be expressed by the customer.

2 Calculated and assumed loads
Solar, transmission, people, machinery or equipment and ventilation. This information is essential as it tells the customer about the basis of the load estimate, especially with regard to the number of people and the amount of machinery or equipment in the space at the time of quotation. If more people or equipment are moved in after installation, the system may not cope. By listing it in your quotation you have a record to check the loads.

3 Equipment
Type, size, location and requirements for condensate disposal.

4 Special requirements
Planning and landlord's permission including noise, indicating who will be responsible for obtaining the necessary permissions. Checking that the incoming electricity supply is adequate. Allowing suitable access, especially during working hours. Who does the 'other trades' work?

Always avoid the use of jargon, as what may be obvious to someone involved in or with the air conditioning industry may not be so to the customer. For example, to state that the equipment is 'to ARI condition A' is meaningless to customers, especially in the UK, as they are unlikely to know of the American Air Conditioning and Refrigeration Institute or its standards. It would be better to set out the actual conditions and relate them to the appropriate British Standard.

The proposals put forward may become part of a contract between the designer and the customer, so avoid leaving loopholes which can lead to later argument.

ELEVEN

Worked examples

Three typical applications for packaged air conditioning

To illustrate the methods for design, equipment selection, etc., in this book, three typical applications are considered in this chapter. They are a mid-floor private office situated in the middle of an office block, a small corner boutique, and a small restaurant situated in what was once a terraced house.

Each example shows the layout of the premises and its dimensions, together with information on the structure, the number of people and any heat-producing equipment.

A load estimate sheet is completed for each example and where a ducted system is chosen the method of duct sizing is shown.

For each example the options and design considerations are discussed, to indicate the manner of achieving a reasonable design for different applications.

Worked example 1: Private office

Figure 11.1 shows the layout of this mid-floor office which is situated in an older office block. This is the only office to be considered at present for air conditioning. Therefore, heat will be conducted into the office from those on either side, those above

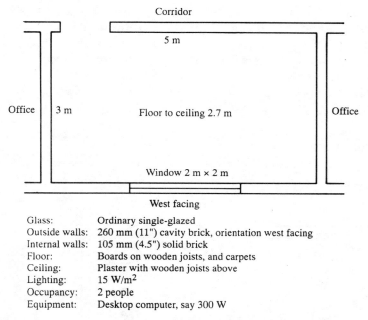

Glass: Ordinary single-glazed
Outside walls: 260 mm (11") cavity brick, orientation west facing
Internal walls: 105 mm (4.5") solid brick
Floor: Boards on wooden joists, and carpets
Ceiling: Plaster with wooden joists above
Lighting: 15 W/m^2
Occupancy: 2 people
Equipment: Desktop computer, say 300 W

Figure 11.1   *Plan and details of mid-floor office (worked example 1)*

and below and from the adjoining corridor. Generally only one person will occupy the office, although occasionally during meetings, etc., there could be one other. A desktop computer is to be considered, but this may not be on all day. Although most people do not smoke, it cannot be guaranteed that the occupant or the visitor will not. Therefore, base the ventilation rate on some smoking unless there is an office policy to ban smoking. If a no smoking ventilation rate is decided upon because of a ban, make sure that this assumption is stated in the proposals.

The office overlooks a courtyard and therefore it can be assumed that the siting of the condensing unit will not cause any problems. However, siting any equipment on the facade or on a roof where it can be seen from street level is subject to planning regulations; therefore the designer or the customer will need to seek permission for siting such equipment. Another consideration will be noise, which can annoy those in the adjacent offices or across the courtyard. Provision must be made to drain the condensate: do not assume that it can be drained to outside as it may drop water on people walking below.

The completed load estimate for the office is shown in Figure 11.2.

CUSTOMER				BUILDING		
NAME PRIVATE OFFICE				TYPE OFFICE		
ADDRESS				LENGTH 5m.		
				WIDTH 3m.		
				HEIGHT 2.7m.		

SURFACE	AREA or No.	FACTOR	SENSIBLE HEAT	LATENT HEAT
GLASS: Sunlit	4	380	1520	
Shade				
WALLS: Sunlit	9.5	25	238	
Shade				
Internal	30	8	240	
ROOF or CEILING	15	8	120	
FLOOR	15	8	120	
LIGHTING	15	15	225	
APPLIANCES			300	
PEOPLE: Sensible Heat	2	91	182	
Latent Heat	2	50		100

VENTILATION – No. People	Requirement	Factor		
2	0.012	4600	110	
2	0.012	9600		230
			3055	330
		TOTAL		3385

Figure 11.2 *Load estimate for mid-floor office (worked example 1)*

As it is a private office it is unlikely that a window unit would be acceptable, so the choice is between a console unit or a split-packaged unit. The problem with the console unit is to provide a suitable hole in the wall above ground level in an existing building. The cost of scaffolding may make this option prohibitive. A split-packaged unit with a wall-hung condensing unit and a low side-wall indoor unit is the best from an installation point of view as well as that concerning noise and design. As the room sensible heat ratio is high, a unit with a high sensible heat ratio is needed. Attention should be paid to the small print in many manufacturers' data: capacities are normally quoted at the worldwide conditions (as BS condition A) which are 27°C db and 19°C wb inside and 35°C db outside. Also, the capacity is normally given with the unit operating at its maximum speed, which is usually too noisy for most commercial applications, except as a boost for start-up.

The load of the office is 3055 W of sensible heat with only 330 W of latent heat, which gives a sensible heat ratio of 0.90. The design temperatures can be assumed as 23°C db and 16.5°C wb inside and 27°C db outside.

The information provided by a typical manufacturer would give a capacity at the higher conditions given above. Assume that the capacity for the nearest suitably sized unit is 3550 W. As the unit is designed for a general application, the sensible heat ratio of this standard unit will be about 0.75. The manufacturer selected in this example also gives information at other conditions and fan speeds. At the conditions which are nearer to those required, where the inside temperature is 21°C db and 15°C wb, with an outside temperature of 27°C db (BS condition C), the capacity will be reduced to 3250 W. At the conditions assumed for the design, the capacity will be slightly higher but not enough to matter in this case. These capacities, however, are with the unit fan at high speed. To arrive at the probable capacity for medium fan speed, a small calculation is necessary. At high speed the air volume is 0.18 m³/s; therefore using the formula:

$$\text{Temp. diff.} = \frac{\text{Sensible heat}}{1200 \times \text{Air vol.}} = \frac{2437}{1200 \times 0.18} = 11.3°C$$

At medium speed the air volume is $0.16\,\text{m}^3/\text{s}$; therefore

$$\text{Sensible heat} = 1200 \times \text{Temp. diff.} \times \text{Air vol.}$$
$$= 1200 \times 11.3 \times 0.16 = 2170\,\text{W}$$

This is only 71% of the sensible capacity required, with a much too low sensible heat ratio.

The manufacturer also gives an option for a high sensible heat ratio. This is done by using the next larger size indoor unit than that normally matched to a particular outdoor unit or condensing unit. This changes the characteristics of the refrigeration circuit and gives a sensible heat of 3337 W with a sensible heat ratio of 0.845 at high speed, and 2936 W with a similar sensible heat ratio at medium speed. This compares very favourably with the sensible heat required and although it will remove more latent heat than required, this will still be within the lower limit of comfort of 35% RH. Had there not been a high sensible heat option, the unit would either have had to be sized to deal with the sensible heat required or for the total cooling load. The first would remove too much latent heat, resulting in a very dry atmosphere, while the second would make the unit undersized and unable to cope with the sensible load.

Worked example 2: Corner boutique

Figure 11.3 shows the layout of a typically privately owned boutique in one of the more select areas of a town. It is competing with the national chain stores by providing service and exclusivity. The clientele are unlikely to be casual and will tend to remain for a time to browse, try on dresses and to discuss style with the staff. There are usually three staff on duty throughout the day.

The completed load estimate for the shop is shown in Figure 11.4.

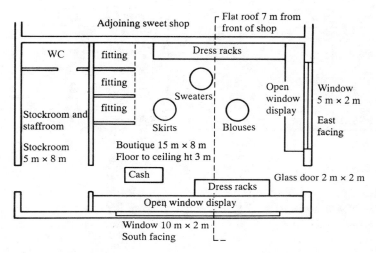

Figure 11.3 *Plan and details of boutique (worked example 2)*

Glass: Single-glazed plate
Outside walls: 260 mm (11") cavity brick
Party walls: 220 mm (9") solid brick
Floor: Concrete
Ceiling: Concrete
Lighting: General, 20 W/m² Display, 20 spotlights at 150 W each
Occupancy: Average 12 people

There is no space for condensing units either on the roof or at the back which leads directly onto a service road. Individual units with their own wall-hung condensing units would be difficult to accommodate.

The best solution is a single packaged ceiling void unit, suspended at high level in the stock room. There is sufficient ceiling height to accommodate the unit without encroaching into the occupied space.

With all packaged air conditioning equipment there is a limited range of options available for a particular model and those for the ceiling void unit less than most. However, one is available that has a total capacity which is near the design but has a sensible heat

CUSTOMER		BUILDING	
NAME CORNER BOUTIQUE		TYPE SHOP.	
ADDRESS		LENGTH 15 m.	
		WIDTH 8 m.	
		HEIGHT 3 m.	

SURFACE	AREA or No.	FACTOR	SENSIBLE HEAT	LATENT HEAT
GLASS: Sunlit	20	380	7600	
Shade	14	95	1330	
WALLS: Sunlit	25	30	750	
Shade	10	11	110	
Internal	69	8	552	
ROOF or CEILING	56	15	840	
	64	8	512	
FLOOR	120	8	960	
LIGHTING	120	20	2400	
	20	150	3000	
APPLIANCES	–	–	1200	
PEOPLE: Sensible Heat	12	100		720
Latent Heat	12	60		

VENTILATION – No. People	Requirement	Factor		
12	0.008	4600	442	
12.	0.008	9600		922
			19698	1642.
		TOTAL	21340.	

Figure 11.4 *Load estimate for boutique (worked example 2)*

ratio and air volumes which are less than required. Even with these differences in criteria the conditions will be within the acceptable limits for comfort, with the temperature being slightly higher and the humidity slightly lower.

Having selected the unit, the air volume used for duct sizing will be that declared by the manufacturer in his data, not the volume based on the design.

This type of shop usually has a ban on smoking and there will only be a limited number of people; therefore sufficient ventilation air should infiltrate naturally through doors and windows, so there is no need to make separate provision for a fresh air inlet into the duct system.

Having decided to use a ceiling void unit, it follows that the system will be ducted. The ducts can either run at high level across the shop over the fitting cubicles or down the length of one side of the shop. The run will depend on the diffusers selected, but there are other aspects to be considered. For example, to discharge air with sufficient velocity to throw towards the shopfront 15 m away without creating cold down draughts, especially with such a limited ceiling height, is very difficult. Therefore, a better solution will be a duct down the length of the shop. The best side for the duct to run will be over the dress racks on the party wall. The throw will be towards the open display window where customers have limited access. In this situation, if there were a tendency for the air to drop into the occupied space at the end of its throw, it is unlikely to cause problems. Also, the shop customers will not be static and therefore can move away from any uncomfortable areas. The two places where care needs to be taken will be around the cash desk, where the staff have to work, and the fitting cubicles.

Condenser air intake and discharge can be taken from high level in the back wall to the back of the unit. The supply air can be taken from the side of the unit to the party wall, turn along the wall over the dress racks, serving diffusers blowing across the shop. The return air can be from a grille over the fitting cubicles ducted into the front of the unit. The condensate can be drained into the toilet basin drain.

The size and number of diffusers will, in practice, be a compromise between the various selection parameters, such as throw and drop. From the layout, the throw cannot be more than 7 m and the mounting height must be less than 3 m or with a drop of more than say 1 m. A number of diffusers are required to limit each to an acceptable spread. By considering these limitations, the acceptable noise level of the application and the air volume that has to be supplied to the space, the number and size of the diffusers can be determined. In this example there are six each, sized 600 mm × 200 mm and having a resistance of 7 Pa. Knowing the general layout, the number of diffusers and the air flow through each diffuser, the number of branches and duct sections can be determined and sized (Table 11.1).

Had there been room for the condensing units and such units would not have caused any noise problems, then the cheapest and most economic design would have been one using three indoor units mounted under the ceiling or at high level on the wall. There could be a minor problem related to the condensate removal, but this could be overcome by condensate pumps fitted in each indoor unit.

Worked example 3: Restaurant

Figure 11.5 shows a typical small restaurant, of the type seen throughout the country. The premises housing the restaurant were once a private terraced house in the Edwardian era, converted initially to a shop where the front was extended into the old front yard and a further extension at a later time into the back yard for a store room or a kitchen. The upstairs is now a flat, with the front extension forming a balcony.

The condensing units can be installed on either the front balcony or the roof of the back extension, depending on the requirements of the owner and the occupier of the flat.

Although the restaurant seats 40 people (40 covers) it will be very rare that the tables are fully occupied and certainly not for a long period of time. Therefore take an average number, which in this example can be 35 people. Discussion with the manager will give a better indication of the average. In most restaurants there is only limited smoking even if there is no ban, and in some it is allowed only in certain areas. This makes the ventilation requirements less critical than when the majority of people smoked. The ventilation rate in most restaurants should be based on some smoking rather than heavy smoking.

Table 11.1   Duct sizing for boutique (worked example 2)

Section	Ref.	Vol.	Dia.	Rect.	Lgth	Rest/m	Rest	Vel.	Vel. pr.
Ret. air	X	1.26	525	500 × 450	1	0.6	0.6	5.6	18.82
Mn duct	AB	1.26	525	500 × 450	7	0.6	4.2	5.6	18.82
Mn duct	BC	1.05	500	450 × 450	2	0.6	1.2	5.5	18.15
Mn duct	CD	0.84	450	400 × 450	2	0.6	1.2	5.0	15.00
Mn duct	DE	0.63	420	350 × 450	2	0.6	1.2	4.7	13.25
Mn duct	EF	0.42	350	250 × 450	2	0.6	1.2	4.25	11.00
Branch	FG	0.21	275	200 × 450	2	0.6	1.2	3.5	7.35
Branch	FH	0.21							
Branch	EI	0.21							
Branch	DJ	0.21	These branches not on the index circuit						
Branch	CK	0.21							
Branch	BL	0.21							

Total straight duct resistance   10.8 Pa

Selecting K factors for bends:
Bends at AB: height 450 mm, width 500 mm, radius 250 mm

Ratio $\dfrac{h}{w}=\dfrac{450}{500}=0.9$; ratio $\dfrac{r}{w}=\dfrac{250}{500}=0.5$; $K=0.35$

Bend at FG: height 450 mm, width 200 mm, radius 200 mm

Ratio $\dfrac{b}{w}=\dfrac{450}{200}=1.25$; ratio $\dfrac{r}{w}=\dfrac{200}{200}=1.0$; $K=0.22$

Section	ref.	Fitting	K factor	Vel. pr.	Resistance
	X	Grille			11.00
	AB	2 bends	0.35	18.82	6.60
	BC	Change	0.04	18.15	0.70
	CD	Change	0.04	15.00	0.60
	DE	Change	0.04	13.25	0.50
	EF	Change	0.04	11.00	0.40
	FG	Change	0.04	7.35	0.30
		Bend	0.22	7.25	1.60
		Diffuser			7.00

Total fittings resistance   28.70

Total resistance   39.50

Add for dirty filter (50 Pa)   89.50

Unit should have an air volume of 1.26 m³ against an external static resistance of 89.50 Pa which is within the range of the ceiling unit selected.

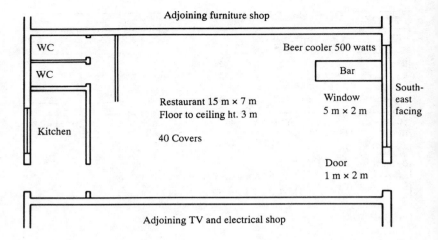

Glass: Single-glazed plate
Outside walls: 220 mm (9") solid brick
Party walls: 220 mm (9") solid brick
Floor: Boards on wooden joists, and carpet
Ceiling: Plaster with wooden joists, above
Lighting: 20 W/m^2
Equipment: Beer cooler 500 W

Figure 11.5 *Plan and details of restaurant (worked example 3)*

A completed load estimate for the restaurant is shown in Figure 11.6.

In a restaurant it is the waiters who can move around and the customers who are static; therefore care must be taken to select and position the indoor units so that they will not drop cold air onto a table, nor be so noisy that they make conversation difficult. It is better to have a number of small units than fewer large units. Depending on the manufacturer, three large units, four smaller units or two multisplit units with four indoor units will cater for the load in the restaurant. The multisplit option would mean that only two condensing or outdoor units need be accommodated which may make siting easier.

CUSTOMER		BUILDING	
NAME RESTAURANT.		TYPE RESTAURANT.	
ADDRESS		LENGTH 15m.	
		WIDTH 7m.	
		HEIGHT 3m.	

SURFACE	AREA or No.	FACTOR	SENSIBLE HEAT	LATENT HEAT
GLASS: Sunlit	12	440	5280.	
Shade				
WALLS: Sunlit	9	22	198	
Shade	111	8	888	
Internal				
ROOF or CEILING	105	8	840	
FLOOR	105	8	840	
LIGHTING	105	20	2100	
APPLIANCES	—		500	
PEOPLE: Sensible Heat	35	110	3850.	
Latent Heat	35	70		2450
VENTILATION – No. People	Requirement	Factor		
35	0.012	4600	1932	
35.	0.012	9600		4032
			16428	6482.
		TOTAL	22910	

Figure 11.6 *Load estimate for restaurant (worked example 3)*

None of these solutions would deal with the fresh or outside ventilation required. To provide for this, the ventilation requirement through the system will necessitate a ducted layout. It is probable, however, that the restaurant has already tried to overcome its overheating problems by installing window fans. Although not the ideal solution, because the air through these fans will not be filtered; it is still an effective and economic way of solving the problem. Also, if smoking is seen as a problem, then this could be minimized by the use of separate electrostatic filters which can remove the smoke from the air rather than dealing with the problem through the cooling and dehumidifying system.

There is a way to reduce the calculated load without creating too many problems, but it must be discussed with the restaurant owners, making sure they understand what is happening and the limitations it imposes. The main load of a restaurant occurs at lunch-time and in the early evening. At both times there is the load from the sun, etc., as well as the internal loads from people and food to consider. Later in the evening the effect of the sun is less and the outside air temperature will be decreasing.

The type of restaurant which usually considers air conditioning will generally open at about 12.30 pm, close at 3.00 pm, then reopen at 7.00 pm until about 11.30 pm. If the restaurant is precooled below comfort levels by adjusting the thermostat to say 19°C, then the structure becomes a form of cool storage. As there are few people in the morning when the staff first arrive, a smaller capacity plant than calculated would be sufficient for this precooling. At opening, the temperature is slightly below that necessary for comfort, but customers coming into an empty restaurant will accept the coolness. As the lunch-time progresses and the load increases due to the numbers of people, the temperature rises because the plant cannot cope. At about 1.30 pm the temperature may have risen to 22°C and by 2.30 pm to about 24°C. By this time some customers will be leaving, thereby reducing the load. Also, when entering a crowded restaurant it is expected to be warmer than average.

When the restaurant closes at 3.00 pm the temperature has either been maintained at about 24°C or maybe less. The air conditioning plant then begins to precool the empty restaurant for the evening. By operating the plant this way, to precool the structure, the capacity can be reduced. For example, the number of units can be reduced from four to three. Even if the owner, having accepted operating the plant this way to reduce capital costs, finds it to be unsatisfactory, an additional unit can easily be added. It should be emphasized that the means of crudely creating cool storage to reduce the load should only be suggested if the owners have to reduce the capital costs, and they should always be made aware of the compromise and the implications it has on the running of the system.

Appendix I

Glossary of air conditioning terms

Absorption refrigeration system   Where the evaporation of the refrigerant is assisted by an absorber (lithium bromide or ammonia). The diluted water/absorber mixture is reconcentrated by heat.

Air conditioner   An encased assembly or assemblies designed as a unit to provide delivery of conditioned air to an enclosed space (a room for instance) or zone. It includes an electrically operated refrigeration system for cooling and possibly dehumidifying the air. It may have means for heating, circulating, cleaning and humidifying the air.

Air conditioner – close control   An air conditioner to satisfy the requirements of the process carried out in the air conditioned room.

Air conditioner – comfort   An air conditioner to satisfy the requirements of the occupants of the air conditioned room.

Air conditioner – single duct   An air conditioner for comfort cooling in which the condenser air intake is introduced from the conditioned space and discharged outside this space.

Air conditioner – spot   An air conditioner that cools a zone within a space. The condenser air is introduced from that space and discharged back into the same space.

Apparatus dew-point   The theoretical coil surface temperature, assuming that the coil is 100% efficient.

Bypass factor   The ratio of air which is cooled to the coil surface temperature to that which bypasses the cooling effect of the coil (a measure of the coil effectiveness)

Capacity (cooling)   The heat removed by the air conditioner from the evaporator or the heat pump when operating in the cooling mode from the indoor heat exchanger.
Capacity (heating)   The usable heat given off by the heat pump when operating in the heating mode from the indoor heat exchanger (for air conditioners, the heating is provided from a separate source).
Centrifugal compressor   A compressor where the compressing is by the centrifugal action of a rotor.
Check or non-return valve   A valve which allows the fluid to flow in one direction only.
Clo factor   An indicator of the insulating effect of clothing.
Coanda effect   The tendency of an airstream to adhere to the ceiling because of surface tension.
Coefficient of performance (COP)   The ratio indicating the effectiveness of a heat pump when heating by dividing the heating capacity by the effective power input.
Compressor   A device to increase the pressure of a refrigerant gas within a refrigeration system.
Condensate   The water condensed on the cooling coil as the air temperature falls below its dew-point.
Condenser   The heat-rejecting coil through which the high-pressure refrigerant gas is condensed to a high-pressure liquid.
Condensing pressure   The refrigerant pressure at the condenser or in the high-pressure side of a refrigeration system.
Condensing temperature   The temperature equivalent to the condensing pressure.

Defrost   The process of removing ice or frost from the surface of a cooling coil.
Dehumidification – comfort   Dehumidification to reduce the humidity within a space to a level to satisfy the requirements of the occupants.

Dehumidification – heat recovery   Dehumidification where the latent and sensible heat removed from the space, together with the compressor heat, is reused in another application rather than rejected outside to waste.

Dehumidification – process   Dehumidification to reduce the humidity within a space to a level necessary for the process or the storage of goods and/or materials or the drying out of the building's fabric.

Dehumidification capacity   The measure of the net removal of moisture by the unit from its surrounding atmosphere (litres/hour).

Dehumidification efficiency ratio (DER)   The ratio of dehumidification capacity to the average power input of the dehumidifier measured over a defined interval of time.

Dehumidifier   An encased assembly designed to remove moisture from its surrounding atmosphere. It includes an electrically operated refrigeration system and a means to recirculate air. It also includes a drain arrangement for collecting and storing and/or disposing of the condensate.

Dew-point   The temperature at which the air is saturated and any further cooling results in condensation of water from the air.

Diffuser   A supply terminal device to ensure the correct entrainment of room air with supply air.

Direct expansion   A refrigerating system where the evaporator is in direct contact with the air.

Direct solar radiation   The radiation received by a surface which is in the direct path of the sun's rays.

Discharge line   The refrigerant pipework joining the compressor and the condenser.

Drop   The amount a horizontal airstream discharging from a diffuser drops towards the end of its throw.

Dry bulb temperature   The temperature indicated by a dry sensing element such as a mercury in glass thermometer.

Effective power   The power input to a unit for the compressor, evaporator and condenser fans, controls and for defrost.

Energy efficiency ratio   The ratio obtained by dividing the cooling capacity by the effective power input of a unit which indicates its effectiveness.

Enthalpy   The total heat (sensible and latent) per unit mass above a set datum.

Entrainment   The induced flow of room air (secondary air) by the supply from the air outlet (primary air).

Equivalent temperature difference   A method of assessing the solar heat gain through a wall or roof by assigning an assumed temperature difference which, when used in conjunction with the 'U' value and area, gives the same value of heat gain.

Evaporating pressure   The refrigerant pressure at the evaporator or low-pressure side of the refrigerant system.

Evaporating temperature   The temperature equivalent to the evaporating pressure.

Evaporator   The heat-absorbing coil through which the low-pressure refrigerant is evaporated into a vapour or gas.

Expansion valve   The valve controlling refrigerant flow through the evaporator and where the refrigerant changes from a high pressure to a low pressure.

Head pressure control   A valve which responds to falling pressure, usually fitted to the outlet of the condenser coil.

Heat of compression   The heat added to the refrigerant through the work done by the compressor.

Heat of rejection or condensation   The heat rejected by the refrigerant through the condenser.

Heat pump   A device which takes heat at a certain temperature and releases it at a higher temperature. When operated to provide heat (e.g. for space heating or water heating), the heat pump is said to operate in the heating mode; when operated to remove heat (e.g. for space cooling), it is said to be operating in the cooling mode.

Hot gas bypass   Where some of the refrigerant gas from the compressor discharge bypasses the condenser and is used to defrost the evaporator.

Hot gas line See Discharge line.
Humidistat A control which senses the relative humidity to control the unit or system.

Indirect solar radiation The radiation received by a surface from the diffuse radiation of the sun caused by dust, water droplets, etc.
Initial velocity The velocity of the air when leaving the diffuser or terminal device.
Inverter control An electronic device which controls the speed of compressors so that the capacity of the system can be varied to match the load of the application by altering the frequency of the standard electricity supply. The device changes the standard a.c. supply to a d.c. supply, then to a variable-frequency a.c. supply.

Latent heat The heat added or removed during a change of state and where the temperature remains constant.
Latent heat of fusion The amount of heat added or removed during a change in state from solid to liquid.
Latent heat of vaporization The amount of heat added or removed during a change in state from liquid to vapour or gas.
Liquid line The refrigerant pipework joining the condenser to the expansion device.
Low ambient control A control to match the heat transfer through the condenser with the load at low external temperature conditions. Usually a speed control on the condenser fan(s).

Micron A measure of particle size in respect to filtration, equivalent to one-thousandth of a millimetre.
Moisture content The actual amount of moisture present in a unit mass of air.
Multisplit system A split-packaged system which can change over to provide either heating or cooling from multiple indoor units, but where only one expansion device controls the flow to all the indoor units.
Multizone heat-recovery system A split-packaged system which can provide simultaneous heating and cooling as required by

multiple indoor units, each of which is controlled by its own expansion device. Heat is recovered and transferred by either two or three pipes between individual indoor units or the outdoor unit.

Multizone modular system   A split-packaged system which can change over to provide either heating or cooling from multiple indoor units, each of which is controlled by its own expansion device.

Occupied zone   That part of the room or space where human activity occurs, usually up to 2.5 m above floor level.

Off-coil temperature   The temperature of the air leaving the coil.

On-coil temperature   The temperature of the air entering the coil.

Percentage saturation   The ratio of moisture content of the air compared with the saturated moisture content at a specific temperature.

Precharged lines   Refrigerant pipework charged at the factory with the required amount of refrigerant for use with a split system.

Primary air   The air delivered to the room through the diffuser or terminal device.

Quick couplings   The block valves or connectors used with precharged lines to retain the refrigerant.

Rated conditions   The standardized conditions for the determination of the data which are characteristic for the air conditioner, heat pump or dehumidifier.

Rating   The capacity of the air conditioner, heat pump or dehumidifier at rated conditions.

Reciprocating compressor   A compressor where the compression is by the reciprocating effect of pistons in a cylinder.

Refrigerant   A fluid which accepts heat at low temperature and discharges it at high temperature within acceptable operating pressures.

Refrigerant charge   The weight or mass of refrigerant needed by a system to transfer the heat capacity of the unit.

Refrigeration effect The amount of heat absorbed by the refrigerant through the evaporator.

Refrigeration system – absorption A refrigeration system in which the refrigerant vapour is absorbed and subsequently expelled by the application of heat.

Refrigeration system – vapour compression A refrigeration system where the refrigerant vapour from the evaporator is compressed then cooled in a condenser before returning to the evaporator.

Relative humidity (RH) The ratio of the water vapour pressure compared with the saturated pressure at a specific temperature.

Reversing valve A valve which changes the direction of the refrigerant discharge from the compressor to the coils and the return from the coils, so that each coil can operate as a condenser or evaporator as required.

Rotary compressor A compressor where the compression is by the rotary action of an eccentric cam within a cylinder.

Saturated air When the air is retaining as much moisture as it can hold at a specific temperature, i.e. 100% relative humidity.

Screw compressor A compressor where the compression is by the contra-rotation of twin helical screws.

Scroll compressor A compressor where the compression is by the action of an eccentric scroll plate within a fixed scroll plate.

Secondary air The air induced into the primary airstream from the room or space.

Sensible heat Heat that affects the temperature of a substance without changing the state.

Sensible heat ratio The ratio of total heat to sensible heat.

Sight glass A component in a refrigeration system which allows the refrigerant flow to be seen.

Single-packaged unit A self-contained unit, factory assembled, with all the components in a single casing and operated as a discrete functional unit.

Specific humidity The ratio of the mass of water vapour in the air to the mass of humidified air.

Specific volume   The volume of air per unit mass.

Split-packaged unit   A factory-assembled unit with the components in two or more casings, but operated as a discrete functional unit.

Split-packaged unit with a single indoor unit   A split-packaged unit where one outdoor unit is matched to one indoor unit.

Split system with multiple indoor units   A split-packaged unit where one outdoor unit is matched to two or more individual indoor units all of which are part of a common refrigeration circuit.

Spread   The amount the primary air spreads when reaching terminal velocity.

Static pressure   The pressure available from a unit to overcome the resistance to air flow in the system.

Strainer   A component in a refrigeration system to remove small particles and contaminants from the system.

Suction line   The refrigerant pipework joining the evaporator to the compressor.

Sun path diagram   A diagram which indicates the azimuth and altitude of the sun at different times of the day and year for a specific latitude.

Terminal velocity   The velocity at which the primary air is said to be approaching that of the natural convection currents in the room or space.

Thermal time lag   The difference between the time a load appears to the time it becomes a load on the room or space.

Thermostat   A control which senses the temperature to control the unit or system.

Throw   The distance the primary air reaches from the diffuser or inlet when its velocity is at the terminal velocity.

Ton of refrigeration   A unit of refrigeration used in the USA based on the cooling effect of one American ton (2000 lb) of ice melting over 24 hours (equivalent to 3.5 kW).

Total heat   The total of both sensible and latent heat.

Vapour compression   A refrigeration system where the refrigerant vapour from the evaporator is compressed, then cooled in a condenser before returning to the evaporator.

Velocity pressure   The pressure equivalent to the fluid velocity.

Wet bulb temperature   The depressed temperature sensed by a sensing element which is kept wet and therefore reduced by the evaporation of moisture from the element. The magnitude of the depression compared with the dry bulb indicates the relative humidity of the air.

Work done   The energy imparted to the refrigerant vapour by the compressor during compression.

Appendix 2

Bibliography

Chartered Institution of Building Service Engineers (1973) *CIBSE Guide*, London.

Electricity Association (1980) *Air conditioning: what it is and what it can do for you*, London.

Electricity Association (1988) *Electric heat pumps*, London.

Electricity Association (1988) *Ventilation in commercial buildings*, London.

Faber, O. and Kell, J.R. (1974) *Heating and Air Conditioning of Buildings*, Butterworth-Heinemann, Oxford.

Heap, R.D. (1979) *Heat Pumps*, E. & F.N. Spon, London.

Heating and Ventilation Contractors Association (1980) *Guide to good practice: Ductwork*, London.

Heating and Ventilation Contractors Association (1992) *Guide to good practice: Unit air conditioning*, London.

Jones, W.P. (1973) *Air Conditioning Engineering*, E. Arnold, London.

For further reading, also consult the various technical information sheets published by the Electricity Association (London) between 1980 and 1990.

Index

The
Whale Rider

ALSO BY WITI IHIMAERA:

NOVELS
Tangi
Whanau
The Matriarch
Bulibasha, King of the Gypsies
Nights in the Gardens of Spain
The Dream Swimmer
The Uncle's Story

SHORT STORY COLLECTIONS
Pounamu Pounamu
The New Net Goes Fishing
Dear Miss Mansfield
Kingfisher Come Home

OTHER COLLECTIONS (EDITOR)
Into the World of Light
Te Ao Marama, vols. 1–5
Vision Aotearoa
Mataora, the Living Face
Growing Up Maori
Te Ata: Māori Art from the East Coast, New Zealand

NON-FICTION
The Legendary Land
Land, Sea, Sky
Masks and Mirror
Maori
On Top Down Under
This Is New Zealand

PLAYS
Woman Far Walking

CHILDREN'S BOOKS
The Little Kowhai Tree

The
Whale Rider

Witi Ihimaera

REED

For Jessica Kiri and Olivia Ata,
the best girls in the whole wide world

Reed Publishing
Te Karuhi tā tāpui o Reed (Aotearoa) (NZ) Ltd

Established in 1907, Reed is New Zealand's largest
book publisher, with over 300 titles in print.

For details on all these books visit our website:
www.reed.co.nz

Published by Reed Books, a division of Reed Publishing (NZ) Ltd,
39 Rawene Rd, Birkenhead, Auckland.
Associated companies, branches and representatives throughout the world.

ISBN 0 7900 0869 6
The author asserts his moral rights in this work.

© 1987 Witi Ihimaera
First published 1987
New edition 1992; reprinted 1993, 1996, 1999, 2001
This edition 2002; reprinted 2003 (six times), 2004 (three times)

Edited by Peter Dowling
Text design by Graeme Leather
Cover design by Serena Kearns
Photography by Kirsty Griffin, courtesy of South Pacific Pictures

Printed in New Zealand

This story is set in Whangara, on the East Coast of New Zealand, where Paikea is the tipuna ancestor. However, the story, people and events described are entirely fictional and have not been based on any people in Whangara.

He tohu aroha ki a Whangara me nga uri o Paikea.

Thanks also to Julia Keelan, Caroline Haapu and Hekia Parata for their advice and assistance.

CONTENTS

The Coming of Kahutia Te Rangi

One

In the old days, in the years that have gone before us, the land and sea felt a great emptiness, a yearning. The mountains were like the poutama, the stairway to heaven, and the lush green rainforest was a rippling kakahu of many colours. The sky was iridescent paua, swirling with the kowhaiwhai patterns of wind and clouds; sometimes it reflected the prisms of rainbow or southern aurora. The sea was ever-changing pounamu, shimmering and seamless to the sky. This was the well at the bottom of the world and when you looked into it you felt you could see to the end of forever.

This is not to say that the land and sea were without life, without vivacity. The tuatara, the ancient lizard with its third eye, was sentinel here, unblinking in the hot sun, watching and waiting to the east. The moa browsed in giant wingless herds across the southern island. Within the warm stomach of the rainforests, kiwi, weka and the other birds foraged for huhu and similar succulent insects. The forests were loud with the clatter of tree bark, chatter of cicada and murmur of fish-laden streams. Sometimes the forest grew suddenly quiet and in wet bush could be heard the fairy filigree of patupaiarehe laughter like a sparkling glissando.

The sea, too, teemed with fish but they also seemed to be waiting. They swam in brilliant shoals, like rains of glittering dust, throughout the greenstone depths — hapuku, manga, kahawai, tamure, moki and warehou — herded by mako or mango ururoa. Sometimes from far off a white shape would be seen flying through the sea but it would only be the serene flight of the tarawhai, the stingray with the spike on its tail.

Waiting. Waiting for the seeding. Waiting for the gifting. Waiting for the blessing to come.

Suddenly, looking up at the surface, the fish began to see the dark bellies of the canoes from the east. The first of the Ancients were coming, journeying from their island kingdom beyond the horizon. Then, after a period, canoes were seen to be returning to the east, making long cracks on the surface sheen. The land and the sea sighed with gladness:

We have been found.
The news is being taken back to the place of the Ancients.
Our blessing will come soon.

In that waiting time, earth and sea began to feel the sharp pangs of need, for an end to the yearning. The forests sent sweet perfumes upon the eastern winds and garlands of pohutukawa upon the eastern tides. The sea flashed continuously with flying fish, leaping high to look beyond the horizon and to be the first to announce the coming; in the shallows, the chameleon seahorses pranced at attention. The only reluctant ones were the patupaiarehe who retreated with their silver laughter to caves in glistening waterfalls.

The sun rose and set, rose and set. Then one day, at its noon apex, the first sighting was made. A spume on the horizon. A dark shape rising from the greenstone depths of the ocean, awesome, leviathan, breaching through the surface and hurling itself skyward before falling seaward again. Under water the muted thunder boomed like a great door opening far away, and both sea and land trembled from the impact of that downward plunging.

Suddenly the sea was filled with awesome singing, a song with eternity in it, a karanga to the land:

> *You have called and I have come,*
> *bearing the gift of the Gods.*

The dark shape rising, rising again. A taniwha, gigantic. A tipua. Just as it burst through the sea, a flying fish leaping high in its ecstasy saw water and air streaming like thunderous foam from that noble beast and knew, ah yes, that the time had come. For the sacred sign was on the monster, a swirling moko pattern imprinted on the forehead.

Then the flying fish saw that astride the head, as it broke skyward, was a man. He was wondrous to look upon, the whale rider. The water streamed away from him and he opened his mouth to gasp in the cold air. His eyes were shining with splendour. His body dazzled with diamond spray. Upon that beast he looked like a small tattooed figurine, dark brown, glistening and erect. He seemed, with all his strength, to be pulling the whale into the sky.

Rising, rising. And the man felt the power of the whale as it propelled itself from the sea. He saw far off the land long sought and now found, and he began to fling small spears of mauri seaward and landward on his magnificent journey toward the land.

Some of the mauri in mid flight turned into pigeons which flew into the forests. Others on landing in the sea changed into eels. And the song in the sea drenched the air with ageless music and land and sea opened themselves to him, the gift long waited for: tangata, man. With great gladness and thanksgiving he, the man, cried out to the land.

Karanga mai, karangi mai, karanga mai.

But there was one mauri, so it is told, the last, which, when the whale rider tried to throw it, refused to leave his hand. Try as he might, the mauri would not fly.

So the whale rider uttered a karakia over the wooden spear, saying, 'Let this mauri be planted in the years to come, for there are sufficient mauri already implanted. Let this be the one to flower when the people are troubled and the mauri is most needed.'

And the mauri then leapt from his hands with gladness and soared through the sky. It flew across a thousand years. When it hit the earth it did not change but waited for another hundred and fifty years to pass until it was needed.

The flukes of the whale stroked majestically at the sky.

Hui e, haumi e, *taiki e.*

SPRING

The Force of Destiny

Two

The Valdes Peninsula, Patagonia. Te Whiti Te Ra. The nursery, the cetacean crib. The giant whales had migrated four months earlier from their Antarctic feeding range to mate, calve and rear their young in two large, calm bays. Their leader, the ancient bull whale, together with the elderly female whales, fluted whalesongs of benign magnificence as they watched over the rest of the herd. In that glassy sea known as the Pathway of the Sun, and under the turning splendour of the stars, they waited until the newly born were strong enough for the long voyages ahead.

Watching, the ancient bull whale was swept up in memories of his own birthing. His mother had been savaged by sharks three months later; crying over her in the shallows of Hawaiki, he had been succoured by the golden human who became his master. The human had heard the young whale's distress and had come into the sea, playing a flute. The sound was plangent and sad as he tried to communicate his oneness with the young whale's mourning. Quite without the musician knowing it, the melodic patterns of the flute's phrases imitated the whalesong of comfort. The young whale drew nearer to the human, who cradled him and pressed noses with the orphan in the first mihimihi. When the herd travelled onward, the young whale remained and grew under the tutelage of his master.

The bull whale had become handsome and virile, and he had loved his master. In the early days his master would play the flute and the whale would come to the call. Even in his lumbering years of age the whale would remember his adolescence and his master; at such moments he would send long, undulating songs of mourning through the lambent water. The elderly females would swim to him hastily, for they loved him, and gently in the dappled warmth they would minister to him.

In a welter of sonics, the ancient bull whale would communicate his nostalgia. And then, in the echoing water, he would hear his master's flute. Straight away the whale would cease his feeding and try to leap out of the sea, as he used to when he was younger and able to speed toward his master.

As the years had burgeoned the happiness of those days was like a siren call to the ancient bull whale. But his elderly females were fearful; for them, that rhapsody of adolescence, that song of the flute, seemed only to signify that their leader was turning his thoughts to the dangerous islands to the south-west.

Three

I suppose that if this story has a beginning it is with Kahu. After all, it was Kahu who was there at the end, and it was Kahu's intervention which perhaps saved us all. We always knew there would be such a child, but when Kahu was born, well, we were looking the other way, really. We were over at our Koro's place, me and the boys, having a korero and a party, when the phone rang.

'A *girl*,' Koro Apirana said, disgusted. 'I will have nothing to do with her. She has broken the male line of descent in our whanau. Aue.' He shoved the telephone at our grandmother, Nanny Flowers, saying, 'Here. It's your fault. Your female side was too strong.' Then he pulled on his gumboots and stomped out of the house.

The phone call was from the eldest grandson, my brother Porourangi, who was living in the South Island. His wife, Rehua, had just given birth to the first great-grandchild of the whanau.

'Tena koe, dear,' Nanny Flowers said into the phone. Nanny Flowers was used to Koro Apirana's growly ways, although she threatened to divorce him every second day, and I could tell that it didn't bother her if the baby was a girl or a boy. Her lips were quivering with emotion because she had been waiting for the call from Porourangi all month. Her eyes went sort of cross-eyed, as they always did whenever she was overcome with aroha. 'What's that? He aha? What did you say?'

We began to laugh, me and the boys, and we yelled to Nanny, 'Hey! Old lady! You're supposed to put the phone to your ear so you can hear!' Nanny disliked telephones; most times she was so shaken to hear a voice come out of little holes in the headpiece that she would hold the phone at arm's length. So I went up to her and put the headpiece against her taringa.

Next minute, the tears started rolling down the old lady's face. 'What's that, dear? *Oh*, the poor thing. Oh the *poor* thing. Oh the poor *thing*. Oh. Oh. Oh. Well you tell Rehua that the first is the worst. The others come easier because by then she'll have the hang of it. Yes, dear. I'll tell him. Yes, don't you worry. Yes. All right. Yes, and we love you too.'

She put down the phone. 'Well, Rawiri,' she said to me, 'you and the boys have got a beautiful mokopuna. She must be, because Porourangi said she looks just like me.' We tried not to laugh, because Nanny was no film star. Then, all of a sudden, she put her hands on her hips and made her face grim and went to the front verandah. Far away, down on the beach, old Koro Apirana was putting his rowboat onto the afternoon sea. Whenever he felt angry he would always get on his rowboat and row out into the middle of the ocean to sulk.

'*Hey*,' Nanny Flowers boomed, 'you old paka,' which was the affectionate name she always called our Koro when she wanted him to know she loved him, 'Hey!' But he pretended he didn't hear her, jumped into the rowboat and made out to sea.

Well, that did it. Nanny Flowers got her wild up. 'Think he can get away from me, does he?' she muttered. 'Pae kare. Pae kare.'

By that time, me and the boys were having hysterics. We crowded onto the verandah and watched as Nanny waddled down the beach, yelling her endearments at her koroua. 'You come back here, you old paka.' Well of course he wouldn't, so next thing, the old lady scooted over to *my* dinghy. Before I could protest she gunned the outboard motor and roared off after Koro Apirana. All that afternoon they were yelling at each other. Koro Apirana would row to one location after another in the bay, and Nanny Flowers would start the motor and roar after him to growl at him. You have to hand it to the old lady, she had brains all right, picking a rowboat with a motor in it. In the end, old Koro Apirana just gave up. He had no chance, really, because Nanny Flowers simply tied his boat to hers and pulled him back to the beach, whether he liked it or not.

That was eight years ago, when Kahu was born, but I remember it as if it was yesterday, especially the wrangling that went on between our Koro and Nanny Flowers. The trouble was that Koro Apirana could not reconcile his traditional beliefs about Maori leadership and mana with Kahu's birth. By Maori custom, leadership was hereditary and normally the mantle of mana fell from the eldest son to the eldest son. Except that in this case, there was an eldest daughter.

'She won't be any good to me,' he would mutter. 'No good.

I won't have anything to do with her. That Porourangi better have a son next time.'

In the end, whenever Nanny Flowers brought the subject up, Koro Apirana would compress his lips, cross his arms, turn his back on her and look elsewhere and not at her.

I was in the kitchen once when this happened. Nanny Flowers was making a Maori bread on the big table, and Koro Apirana was pretending not to hear her, so she addressed herself to me.

'Thinks he knows everything,' she muttered, tossing her head in Koro Apirana's direction. *Bang*, went her fists into the dough. 'The old paka. Thinks he knows all about being a chief.' *Slap*, went the bread as she threw it on the table. 'He isn't any chief. I'm his chief,' she emphasised to me and, then, over her shoulder to Koro Apirana, 'and don't you forget it either.' *Squelch*, went her fingers as she dug them into the dough.

'Te mea te mea,' Koro Apirana said. 'Te mea te mea.'

'Don't you mock *me*,' Nanny Flowers responded. *Ouch*, went the bread as she flattened it with her arms. She looked at me grimly and said, 'He *knows* I'm right. He *knows* I'm a descendant of old Muriwai, and *she* was the greatest rangatira of my tribe. Yeah,' and, *Help*, said the dough as she pummelled it and prodded it and stretched it and strangled it, 'and I should have listened to Mum when she told me not to marry him, the old paka,' she said, revving up to her usual climactic pronouncement.

From the corner of my eye I could see Koro Apirana mouthing the words sarcastically to himself.

'But *this* time,' said Nanny Flowers, as she throttled the bread with both hands, 'I'm *really* going to divorce him.'

Koro Apirana raised his eyebrows, pretending to be unconcerned.

'Te mea te mea,' he said. 'Te mea —'

It was then that Nanny Flowers added with a gleam in her eyes, '*And* then I'll go to live with old Waari over the hill.'

I thought to myself, *Uh oh, I better get out of here*, because Koro Apirana had been jealous of old Waari, who had been Nanny Flowers' first boyfriend, for years. No sooner was I out the door when the battle began. *You coward*, said the dough as I ducked.

Four

But that was nothing compared to the fight that they had when Porourangi rang to say he would like to name the baby Kahu.

'What's wrong with Kahu?' Nanny Flowers asked.

'I know your tricks,' Koro Apirana said. 'You've been talking to Porourangi behind my back, egging him on.'

This was true, but Nanny Flowers said, 'Who, me?' She fluttered her eyelids at the old man.

'You think you're smart,' Koro Apirana said, 'but don't think it'll work.'

This time when he went out to the sea to sulk he took *my* dinghy, the one with the motor in it.

'See if I care,' Nanny Flowers said. She had been mean enough, earlier in the day, to siphon out half the petrol so that he couldn't get back. All that afternoon he shouted and waved but she just pretended not to hear. Then Nanny Flowers rowed out to him and said that, really, there was nothing he could do. She had telephoned Porourangi and said that the baby could be named Kahu, after Kahutia Te Rangi.

I could understand, however, why the old man was so against the idea. Not only was Kahutia Te Rangi a man's name

23

but it was also the name of the ancestor of our village. Koro Apirana felt that naming a girl-child after the founder of our whanau was belittling Kahutia Te Rangi's mana, his prestige. From that time onward, whenever Koro Apirana went past the meeting house, he would look up at the figure of Kahutia Te Rangi on the whale and shake his head sorrowfully. Then he would say to Nanny Flowers, 'You stepped out of line, dear, you shouldn't have done it.' To give credit to her, Nanny Flowers did appear penitent.

I guess the trouble was that Nanny Flowers was always 'stepping out of line'. Even though she had married into our tribe she always made constant reference to her ancestor, Muriwai, who had come to New Zealand on the Maataatua canoe. When the canoe approached Whakatane, which is a long way from our village, Muriwai's chieftainly brothers, led by Toroa, went to investigate the land. While they were away, however, the sea began to rise and the current carried the canoe so close to the rocks that Muriwai knew all on board would surely perish. So she chanted special prayers, asking the gods to give her the right and open the way for her to take charge. Then she cried, 'E-i! Tena, kia whakatane ake au i ahau!' *Now I shall make myself a man.* She called out to the crew and ordered them to start paddling quickly, and the canoe was saved in the nick of time.

'If Muriwai hadn't done that,' Nanny used to say, 'the canoe would have been wrecked.' Then she would hold up her arms and say, 'And I am proud that Muriwai's blood flows in my veins.'

'But that doesn't give you the right,' Koro Apirana said to her one night. He was referring, of course, to her agreeing to the naming of Kahu.

Nanny Flowers went up to him and kissed him on the forehead. 'E Koro,' she said softly, 'I have said prayers about it. What's done is done.'

Looking back, I suspect that Nanny Flowers' action only helped to harden Koro Apirana's heart against his first-born great-grandchild. But Nanny was keeping something back from the old man.

'It's not Porourangi who wants to name the girl Kahu,' she told me. 'It's Rehua.' Then she confided to me that there had been complications in the birth of Kahu and, as a result, the delivery had been by Caesarean section. Rehua, weak and frightened after the birth, had wanted to honour her husband by choosing a name from his people, not hers. That way, should she die, at least her first-born child would be linked to her father's people and land. Rehua was from the same tribe as Nanny Flowers and had that same Muriwai blood, so no wonder she got her way with Porourangi.

Then came a third telephone call from Porourangi. Rehua was still in intensive care and Porourangi had to stay with her, but apparently she wanted Kahu's afterbirth, including the birth cord, to be put in the earth on the marae in our village. An auntie of ours would bring the pito back to Gisborne on the plane the next day.

Koro Apirana was steadfast in his opposition to Kahu.

'She is of Porourangi's blood and yours,' Nanny Flowers said to him. 'It is her right to have her pito here on this marae.'

'Then you do it,' Koro Apirana said.

So it was that Nanny Flowers sought my help. The next day was Friday, and she got dressed in her formal black clothes and put a scarf over her grey hair. 'Rawiri, I want you to take me to the town,' she said.

I got a bit worried at that because Nanny wasn't exactly a featherweight, but she seemed so tense. 'Kei te pai,' I said. So I got my motorbike out of the shed, showed her how to sit on the pillion, put my Headhunters jacket on her to keep her warm, and off we roared. As we were going along Wainui Beach some of the other boys joined us. I thought, 'I'll give Nanny Flowers a thrill and do a drag down the main street.'

Well, Nanny just loved it. There she was, being escorted through the Friday crowd like royalty, waving one hand at everybody and holding on tightly with the other. We had to stop at the lights at Peel Street, and the boys and I gunned our motors, just for her. Some of Nanny's old cronies were crossing; when they saw her through the blue smoke, they almost swallowed their false teeth.

'E hika ma,' they said. 'Ko wai tenei?'

She smiled supremely. 'Ko au te Kuini o nga Headhunters.' At that stage I was getting worried about my shock absorbers, but I couldn't help feeling proud of Nanny. Just as we roared off again she poked out her little finger, as if she was having a cup of tea and said, 'Ta ta.'

But when she met Auntie at the airport, Nanny Flowers' mood changed. We were watching from the road when Auntie got off the plane. She started to cry, and then Nanny started to tangi also. They must have been crying for at least ten minutes before our Auntie passed Kahu's pito to Nanny. Then Auntie escorted Nanny over to us and kissed us all and waved good-bye.

'Take me back to Whangara a quiet way,' Nanny asked. 'I don't want people in the town to see me having a tangi.'

So it was that Nanny and I and the boys returned to the village, and Nanny was still grieving.

She said to me, 'Rawiri, you and the boys will have to help me. Your grandfather won't come. You're the men who belong to this marae.'

The night was falling quickly. We followed Nanny as she went back and forth across the marae. She took a quick look around to make sure no-one was watching us. The sea hissed and surged through her words.

'This is where the pito will be placed,' she said, 'in sight of Kahutia Te Rangi, after whom Kahu has been named. May he, the tipua ancestor, always watch over her. And may the sea from whence he came always protect her through life.'

Nanny Flowers began to scoop a hole in the loose soil. As she placed the pito in it, she said a karakia. When she finished, it had grown dark.

She said, 'You boys are the only ones who know where Kahu's pito has been placed. It is your secret and mine. You have become her guardians.'

Nanny led us to a tap to wash our hands and sprinkle ourselves with water. Just as were going through the gate we saw the light go on in Koro Apirana's room, far away. I heard Nanny whisper in the dark, 'Never mind, Kahu. You'll show him when you grow up. You'll fix the old paka.'

I looked back at the spot where Kahu's pito had been placed. At that moment the moon came out and shone full upon the carved figure of Kahutia Te Rangi on his whale. I saw something flying through the air.

Then, far out to sea, I heard a whale sounding. Hui e, haumi e, *taiki e*.

SUMMER

Halcyon's Flight

Five

Uia mai koia whakahuatia ake, Ko wai te whare nei e? Ko Te Kani! Ko wai te tekoteko kei runga? Ko Paikea, ko Paikea! Whakakau Paikea *hei*, Whakakau he tipua *hei*, Whakakau he taniwha *hei*, Ka u a Paikea ki Ahuahu, *pakia*, Kei te whitia koe, ko Kahutia Te Rangi, *aue*, Me awhi o ringa ki te tamahine, A te Whironui, *aue*, Nana i noho Te Roto-o-Tahe, *aue*, *aue*, He Koruru koe, koro e!

*Four hundred leagues from Easter Island. Te Pito o te Whenua.
Diatoms of light shimmered in the cobalt-blue depths of the
Pacific. The herd, sixty strong, led by its ancient leader, was
following the course computed by him in the massive banks of
his memory. The elderly females assisted the younger mothers,
shepherding the new-born in the first journey from the cetacean
crib. Way out in front, on point and in the rear, the young males
kept guard on the horizon. They watched for danger, not from
other creatures of the sea, but from the greatest threat of all —
man. At every sighting they would send their ululation back to
their leader. They had grown to rely on his memory of the
underwater cathedrals where they could take sanctuary, often
for days, until man had passed. Such a huge cathedral lay*

beneath the sea at the place known as the Navel of the Universe.

Yet it had not always been like this, the ancient whale remembered. Once, he had a golden master who had wooed him with flute song. Then his master had used a conch shell to bray his commands to the whale over long distances. As their communication grew so did their understanding and love of each other. Although the young whale had then been almost twelve metres long, his golden master had begun to swim with him in the sea.

Then, one day, his master impetuously mounted him and became the whale rider. In ecstasy the young male had sped out to deep water and, not hearing the cries of fear from his master, had suddenly sounded in a steep accelerated dive, his tail stroking the sky. In that first sounding he had almost killed the one other creature he loved.

Reminiscing like this the ancient bull whale began to cry his grief in sound ribbons of overwhelming sorrow. Nothing that the elderly females could do would stop his sadness. When the younger males reported a man-sighting on the horizon it took all their strength of reasoning to prevent their leader from arrowing out towards the source of danger. Indeed, only after great coaxing were they able to persuade him to lead them to the underwater sanctuary. Even so, they knew with a sense of inevitability that the old one had already begun to sound to the source of his sadness and into the disturbing dreams of his youth.

Six

Three months after Kahu's birth her mother, Rehua, died. Porourangi brought her and Kahu back to our village where the tangi was held. When Rehua's mother asked if she and her people could raise Kahu, Nanny Flowers objected strongly. But Porourangi said, 'Aue,' and Koro Apirana said, 'Hei aha,' and thereby overruled her.

A week later, Rehua's mother took Kahu from us. I was with Nanny Flowers when the taking occurred. Although Porourangi was in tears, Nanny was strangely tranquil. She held Kahu close, a small face like a dolphin, held and kissed her.

'Never mind, boy,' she said to Porourangi. 'Kahu's pito is here. No matter where she may go, she will always return. She will never be lost to us.' Then I marvelled at her wisdom and Rehua's in naming the child in our whakapapa and the joining of her to our whenua.

Our whakapapa, of course, is the genealogy of the people of Te Tai Rawhiti, the people of the East Coast; Te Tai Rawhiti actually means 'the place washed by the eastern tide'. Far away beyond the horizon is Hawaiki, our ancestral island homeland, the place of the Ancients and the Gods, and the other side of the world. In between is the huge seamless marine continent which we call Te Moana Nui a Kiwa, the Great Ocean of Kiwa.

The first of the Ancients and ancestors had come from the east, following the pathways in the ocean made by the morning sun. In our case, our ancestor was Kahutia Te Rangi, who was a high chief in Hawaiki. In those days man had power over the creatures of land and sea, and it was Kahutia Te Rangi who travelled here on the back of a whale. This is why our meeting house has a carving of Kahutia Te Rangi on a whale at the apex. It announces our pride in our ancestor and acknowledges his importance to us.

At the time there were already people, tangata, living in this land, earlier voyagers who had come by canoe. But the land had not been blessed so that it would flower and become fruitful. Other tribes in Aotearoa have their own stories of the high chiefs and priests who then arrived to bless their tribal territories; our blessing was brought by similar chiefs and priests, and Kahutia Te Rangi was one of them. He came riding through the sea, our sea god Kahutia Te Rangi, astride his tipua, and he brought with him the mauri, the life-giving forces which would enable us to live in close communion with the world. The mauri that he brought came from the Houses of Learning called Te Whakaeroero, Te Rawheoro, Rangitane, and Tapere Nui a Whatonga. They were the gifts of those houses in Hawaiki to the new land. They were very special because among other things, they gave instructions on how man might korero with the beasts and creatures of the sea so that all could live in helpful partnership. They taught *oneness*.

Kahutia Te Rangi landed at Ahuahu, just outside our village, in the early hours of the morning. To commemorate his voyage he was given another name, Paikea. At the time of landfall the star Poututerangi was just rising above our sacred mountain, Hikurangi. The landscape reminded Paikea of his birthplace back in Hawaiki so he named his new home Whangara Mai Tawhiti, which we call Whangara for short. All the other places around here are also named after similar headlands and mountains and rivers in Hawaiki — Tawhiti Point, the Waiapu River, and Tihirau Mai Tawhiti.

It was in this land that Paikea's destiny lay. He married the daughter of Te Whironui, and they were fruitful and had many sons and grandsons. And the people lived on the lands around his pa Ranginui, cultivating their kumara and taro gardens in peace and holding fast to the heritage of their ancestors.

Four generations after Paikea, was born the great ancestor Porourangi, after whom my eldest brother is named. Under his leadership the descent lines of all the people of Te Tai Rawhiti were united in what is now known as the Ngati Porou confederation. His younger brother, Tahu Potiki, founded the South Island's Kai Tahu confederation.

Many centuries later, the chieftainship was passed to Koro Apirana and, from him, to my brother Porourangi. Then Porourangi had a daughter whom he named Kahu.

That was eight years ago, when Kahu was born and then taken to live with her mother's people. I doubt if any of us realised how significant she was to become in our lives. When a child is growing up somewhere else you can't see the small tohu, the signs, which mark her out as different, someone who was meant to be. As I have said before, we were all looking somewhere else.

Eight years ago I was sixteen. I'm twenty-four now. The boys and I still kick around and, although some of my girlfriends have tried hard to tempt me away from it, my first love is still my BSA. Once a bikie always a bikie. Looking back, I can truthfully say that Kahu was never forgotten by me and the boys. After all, we were the ones who brought her pito back to the marae, and only we and Nanny Flowers knew where it was buried. We were Kahu's guardians; whenever I was near the place of her pito, I would feel a little tug at my motorbike jacket and a voice saying, 'Hey Uncle Rawiri, don't forget me.' I told Nanny Flowers about it once and her eyes glistened. 'Even though Kahu is a long way from us she's letting us know that she's thinking of us. One of these days she'll come back.'

As it happened, Porourangi went up to get her and bring her back for a holiday the following summer. At that time he had returned from the South Island to live in Whangara but to work in the city. Koro Apirana was secretly pleased with this arrangement because he had been wanting to pass on his knowledge to Porourangi. One of these days my eldest brother will be the big chief. All of a sudden, during a haka practice on the marae, Porourangi looked up at our ancestor Paikea and said to Koro Apirana, 'I am feeling very mokemoke for my daughter.' Koro Apirana didn't say a word, probably hoping that Porourangi would forget his loneliness. Nanny Flowers, however, as quick as a flash, said, 'Oh you poor thing. You better go up and bring her back for a nice holiday with her grandfather.' We knew she was having a sly dig at Koro Apirana. We could also tell that *she* was mokemoke too for the mokopuna who was so far flung away from her.

On Kahu's part, when she first met Koro Apirana, it must have been love at first sight because she dribbled all over him. Porourangi had walked through the door with his daughter and Nanny Flowers, cross-eyed with joy, had grabbed Kahu for a great big hug. Then, before he could say 'No' she put Kahu in Koro Apirana's arms.

'E hika,' Koro Apirana said.

'A little huware never hurt anybody,' Nanny Flowers scoffed.

'That's not the end I'm worried about,' he grumbled, lifting up Kahu's blankets. We had to laugh, because Kahu had done a mimi.

Looking back, I have to say that that first family reunion with Kahu was filled with warmth and aroha. It was surprising how closely Kahu and Koro Apirana resembled each other. She was bald like he was and *she* didn't have any teeth either. The only difference was that she loved him but he didn't love her. He gave her back to Nanny Flowers and she started to cry, reaching for him. But he turned away and walked out of the house.

'Never mind, Kahu,' Nanny Flowers crooned. 'He'll come around.' The trouble was, though, that he never did.

I suppose there were many reasons for Koro Apirana's attitude. For one thing, both he and Nanny Flowers were in their seventies and, although Nanny Flowers still loved grand-children, Koro Apirana was probably tired of them. For another, he was the big chief of the tribe and was perhaps more preoccupied with the many serious issues facing the survival of the Maori people and our land. But most of all, he had not wanted an eldest girl-child in Kahu's generation; he had wanted an eldest boy-child, somebody more appropriate to teach the traditions of the village to. We didn't know it at the time, but he had already begun to look in other families for such a boy-child.

Kahu didn't know this either, so of course, her love for him remained steadfast. Whenever she saw him she would try to sit up and to dribble some more to attract his attention.

'That kid's hungry,' Koro Apirana would say.

'Yeah,' Nanny Flowers would turn to us, 'she's hungry for *him*, the old paka. Hungry for his love, his aroha. Come to think of it, I must get a divorce and find a young husband.' She and all of us would try to win Kahu over to us but, no, the object of her affection remained a bald man with no teeth.

At that time there was still nothing about Kahu which struck us as out of place. But then two small events occurred. The first was when we discovered that Kahu adored the Maori kai. Nanny had given her a spoonful of kanga pirau, fermented corn, and next minute Kahu had eaten the lot. 'This kid's a throwback,' Nanny Flowers said. 'She doesn't like milk or hot drinks, only cold water. She doesn't like sugar, only Maori kai.'

The second event happened one night when Koro Apirana was having a tribal meeting at the house. He had asked all the men to be there, including me and the boys. We crowded into the sitting room and after prayer and a welcome speech, he got down to business. He said he wanted to begin a regular instruction period for the men so that we would be able to

37

learn our history and our customs. Just the men, he added, because men were tapu. Of course the instruction wouldn't be like in the old days, not as strict, but the purpose would be the same: to keep the reo going, and the mana of the iwi. It was important, he said, for us to be so taught. The lessons would be held in the meeting house and would begin the following week.

Naturally we all agreed. Then, in the relaxed atmosphere that always occurs after a serious korero, Koro Apirana told us of his own instruction years ago under the guidance of a tohunga. One story followed another, and we were all enthralled because the instruction had mainly taken the form of tests or challenges which he had to pass: tests of memory, as in remembering long lines of genealogy; tests of dexterity, wisdom, physical and psychological strength. Among them had been a dive into deep water to retrieve a carved stone dropped there by the tohunga.

'There were so many tests,' said Koro Apirana, 'and some of them I did not understand. But I do know the old man had the power to talk to the beasts and creatures of the sea. Aue, we have lost that power now. Finally, near the end of my training, he took me into his nikau hut. He put out his foot and pointing to the big toe, said "Ngaua." So I did, and —'

Suddenly, Koro Apirana broke off. A look of disbelief spread over his face. Trembling, he peered under the table, and so did we. Kahu was there. Somehow she had managed to crawl unobserved into the room. Koro Apirana's toes must have looked juicy to her because there she was, biting on his big toe and making small snarling sounds as she played with

38

it, like a puppy with a bone. Then she looked up at him, and her eyes seemed to say, 'Don't think you're leaving *me* out of this.'

We were laughing when we told Nanny Flowers.

'I don't know what's so funny,' she said sarcastically, 'Kahu could have gotten poisoned. But good on her to take a bite at the old man. Pity she doesn't have any teeth.'

Koro Apirana, however, was not so amused and now I understand why.

Seven

The next time Kahu came to us she was two years old. She came with Porourangi, who had a lovely woman called Ana with him. It looked like they were in love. But Nanny Flowers had eyes only for Kahu.

'Thank goodness,' Nanny Flowers said after she had embraced Kahu, 'you've grown some hair.'

Kahu giggled. She had turned into a bright button-eyed little girl with shining skin. She wanted to know where her grandfather was.

'The old paka,' said Nanny Flowers. 'He's been in Wellington on Maori Council business. But he comes back on the bus tonight. We'll go and pick him up.'

We had to smile, really, because Kahu was so eager to see Koro Apirana. She wriggled and squirmed all the way into town. We bought her a soft drink but she didn't want it, preferring water instead. Then, when the bus arrived and Koro Apirana stepped off with other Council officials, she ran at him with a loud, infectious joy in her voice. I guess we should have expected it, but it was still a surprise to hear her greeting to him. For his part, he stood there thunderstruck, looking for somewhere to hide.

Oh the shame, the embarrassment, as she flung herself into his arms, crying, 'Oh, *Paka*. You home now, you *Paka*. Oh, *Paka*.'

He blamed us all for that, and he tried to persuade Kahu to call him 'Koro', but *Paka* he was, and *Paka* he became forever after.

Being a big chief, Koro Apirana was often called to hui all over the country to represent us. He had the reputation of being stern and tyrannical and because of this many people were afraid of him. 'Huh,' Nanny Flowers used to say, 'they should face *me* and then they'll know all about it.' But me and the boys had a grudging admiration for the old fella. He might not always be fair but he was a good fighter for the Maori people. Our pet name for our Koro was 'Super Maori' and, even now, telephone boxes still remind me of him. We used to joke: 'If you want help at Bastion Point, call Super Maori. If you want a leader for your Land March, just dial Whangara 214K. If you want a man of mana at a Waitangi protest, phone the Maori Man of Steel.' Mind you, he wasn't on our side when we protested against the Springbok Tour but then that just shows you the kind of man he was: his own boss. 'Right or wrong,' Nanny Flowers would add.

The hui that Koro Apirana had attended was about the establishment of Kohanga Reo, or language nests, where young children could learn the reo. The adult version was the wananga, the regular instruction of the kind which Koro Apirana had established a year before in Whangara. Although we weren't that well educated, the boys and I enjoyed the lessons every weekend. It soon became obvious that Kahu did also. She would sneak up to the door of the meeting house and stare in at us.

'Haere atu koe,' Koro Apirana would thunder. Quick as a flash Kahu's head would bob away. But slowly we would see it again, like a spiny kina. I suspect that Kahu overheard more than we thought. I am certain she must have been there when we learnt that man was once able to talk, to communicate, with whales. After all, Paikea must have had to tell his whale where to come.

The whale has always held a special place in the order of things, even before those times of Paikea. That was way back, after the Sky Father and Earth Mother had been separated, when the God children of both parents divided up between themselves the various Kingdoms of the Earth. It was the Lord Tangaroa who took the Kingdom of the Ocean; he was second in rank only to the Lord Tane, the Father of Man and the Forests, and so was established by them the close kinship of man with the inhabitants of the ocean, and of land with sea. This was the first communion.

Then the Lord Tangaroa appointed the triad of Kiwa, Rona and Kaukau to assist his sovereign rule: Kiwa to be guardian of the southern ocean, Rona to help control the tides and Kaukau to aid the welfare of the sea's denizens. To the triad, two other guardians from the Kingdom of the Land, Takaaho and Te Pu-whakahara, brought a special suit: their offspring had been given lakes to live in, but they preferred to roam the freedom of the sea. The suit was accepted, and this was how sharks and whales were granted habitation of the ocean.

From the very beginning the whale was grateful for this release and this was why the whale family, the Wehenga-kauiki, became known as the helpers of men lost at sea. Whenever asked, the whale would attend the call, as long as the mariner possessed the necessary authority and knew the way of talking to whales.

But as the world aged and man grew away from his godliness, he began to lose the mauri of speech with whales, the power of *interlock*. So it was that the knowledge of whale-speaking was given only to a few. One of these was our ancestor, Paikea.

Then came the time when Paikea asked his whale to bring him to our land, far to the south, and it was done.

As for the whale itself, some people say the tipua was trans-formed into an island; viewed from the highway to Tolaga Bay, the island certainly *does* look like a whale breaking through the water.

The years went by, and the descendants of Paikea increased on the land and always paid homage to their ancestor and the whale island. In those days there was still communion with the Gods and a close relationship between land inhabitants and ocean inhabitants. Whenever man wished to cross the border between his kingdom and that of the ocean he would honour Tangaroa by making offerings of seaweed, or fish or birds. And when Tangaroa granted man good fishing, man would return the first fish of the catch to the sea god as acknow-ledgement that his welfare was only by leave of Tangaroa. So it was that ceremonials of respect were employed between man and sea. For instance, fishing was tapu and women therefore did not go out with the men, and fishing grounds became steeped in special rituals to ensure their bounty. And even the shark, in those days, was a helper to man unless man had transgressed a tapu.

Until the time came when man turned on the beast which had been companion to him and the whalekilling began.

That night, after the wananga on the whales, I arrived home to find Nanny Flowers out on the verandah with Kahu in her arms, rocking back and forth, back and forth.

'Rawiri, what happened down there?' she asked, jerking her head at the meeting house. I saw Kahu rubbing small fists against her eyes.

'Nothing,' I answered. 'Why?'

'This kid has been sobbing her heart out,' Nanny Flowers said. She paused. 'Did the old paka growl at her?'

Ever since the wananga had started, Nanny Flowers had been chucking off at Koro Apirana. While she agreed that the instruction should take place, she couldn't help feeling affronted about the exclusion of women. 'Them's the rules,' Koro Apirana had told her. 'I know, but rules are made to be broken,' she had replied in a huff. So, every first Saturday of the month, she would start to play up and pick on Koro Apirana. 'Te mea te mea,' he would say. 'Te mea te mea.'

'He didn't growl at Kahu any more than usual,' I answered. 'he just doesn't like her hanging around when we have the wananga, that's all.'

Nanny Flowers compressed her lips. I could tell that rebellion was ready to boil over inside her. Then she said to me, 'Well you take this kid with you somewhere because I'm going to have a word with Koro Api when he gets back, the old paka.'

I must admit that I was brassed off, having Kahu shoved at me like that. I was planning on taking my darling Cheryl Marie to the movies. So I phoned her up to explain that I had to look after a baby.

'Oh yeah,' my darling said sarcastically. 'And I suppose she's not five foot two with eyes of blue.' Cheryl was jealous of my other darling, Rhonda Anne.

'No,' I said. 'My baby is *you*. Eyes of brown and lives in town.'

Would you believe it, my darling hung up on me? So what else could I do except take Kahu to the movies instead. The boys laughed when I zoomed up to the Majestic with my substitute 'date' under my leather jacket, but the girls loved her. 'Oh isn't she gorgeous? Isn't she sweet?' Yuk. I could see a mile off that the girls were also assessing whether I had now become marrying material. No *way*.

The movie had already started. Children weren't supposed to see it, but the darkness made it easier to sneak Kahu in. What I hadn't realised, however, was that the main feature was about a whale being hunted through Antarctic waters. Everything was fine, really, for most of the film, because Kahu soon fell asleep. Having her curled up so close to me made me feel protective, like a father, I guess, and I think my bonding to her was confirmed that night. I felt I should look after her till the world ended; every now and then, I would open my jacket and sneak a look at her tiny face, so wan in the light of the flickering film. And a lump would come to my throat and I would think to myself, 'No, Kahu, I won't forget you, ever.'

Then the final tragedy of the movie began. The whale, wounded, was dying in its own blood. The soundtrack was suddenly filled with the sound of the whale in its death throes: long, echoing, sighing phrases which must have been recorded from real whales. The sound was strange and utterly sad. No wonder when I looked at Kahu she had woken from sleep, and tears were again tracking down her face. Not even a lolly would help to pacify her.

Nanny Flowers and Koro Apirana had finished their argument by the time I returned home, but the atmosphere was as frozen as the Antarctic wasteland in the film.

'He's sleeping in the bunkhouse with you tonight,' Nanny Flowers told me, jerking her head at Koro Apirana. 'I've had enough of him. Divorce tomorrow, I mean it this time.' Then she remembered something and after taking Kahu from me, screwed my ears. *Ouch*. 'And that'll teach you to take my mokopuna gallivanting all over the place. I've been scared to death. Where'd you go?'

'To the movies.'

'To a *picture*?' *Bang* came her open hand over my head. 'And *then* where!'

'Down the beach.'

'The *beach*?' I ducked her hand (Ha ha, ha ha, you missed me, you missed —) but *kick* came her foot to my umu. 'Don't you do that again!' She hugged Kahu tightly and took her into hers and Koro Apirana's bedroom and *slam* went the door.

I thought of my darling, Cheryl Marie. 'Looks like both of us lucked out tonight,' I said to Koro Apirana.

Half way through the night I suddenly remembered something. I tried to wake Koro Apirana, snoring beside me, but he only tried to snuggle up to me, saying 'Putiputi, taku wahine . . .' So I edged away from him quickly and sat there, staring through the window at the glowing moon.

I had wanted to tell Koro Apirana that on our way back from the movie, the boys and I had gone up to the Point at Sponge Bay. The sea had looked like crinkled silver foil smoothed right out to the edge of the sky.

'Hey!' one of the boys had said, pointing. 'There's orca.'

It had been uncanny, really, seeing those killer whales slicing stealthily through the sea, uncanny and disturbing as a dream.

Even more strange, though, was that Kahu had begun to make eerie sounds in her throat. I swear that those long lamenting sighs of hers were exactly the same as I had heard in the movie theatre. It sounded as if she was warning them.

The orca suddenly dived.

Hui e, haumi e, *taiki e.*

Eight

The following summer, when Kahu was three, was dry and dusty on the Coast. Koro Apirana was concerned about our drinking water and was considering at one point bringing it in by road tanker. One of the boys suggested that the sweetest water was DB light brown and that the hotel up at Tatapouri would be happy to deliver it free. Another of the boys added that we'd have to escort it to Whangara because, for sure, someone would want to do a Burt Reynolds and hijack it.

Into all this rough and tumble of our lives, Kahu brought a special radiance. Koro Apirana was as grumpy with her as ever but, now that Porourangi was home, and now that the wananga sessions were attracting young boys for him to teach, he seemed to bear less of a grudge against her for being a girl and the eldest grandchild.

'Don't blame Kahu,' Nanny Flowers used to growl. 'If your blood can't beat my Muriwai blood that's your lookout.'

'Te mea te mea,' Koro Apirana would reply. 'Te mea te mea.'

In particular, Koro Apirana had discovered three sons from mana bloodlines to whom he hoped to pass the mantle of knowledge. And from the corner of his eye, he could see that Porourangi and his new whaiaipo, Ana, were growing very fond of each other. Now *she* didn't have any Muriwai blood so, you never knew, Porourangi might come up with a son yet.

Under these conditions, the love which Kahu received from Koro Apirana was the sort that dropped off the edge of the table, like breadcrumbs after everybody else has had a big feed. But Kahu didn't seem to mind. She ran into Koro Apirana's arms whenever he had time for her and took whatever he was able to give. If he had told her he loved dogs I'm sure she would have barked, 'Woof woof'. That's how much she loved him.

Summer is always shearing season for us and that summer the boys and I got a contract to shear for the local farmers around the Coast. On the first few mornings when Kahu was at home I would see her staring at us over the windowsill as we left. Her eyes seemed to say, 'Hey, don't forget about me, Uncle Rawiri.' So one morning I made her life happy.

'I think I'll take Kahu to the shed with me,' I said to Nanny Flowers.

'E hara,' Nanny Flowers said. 'She'll drown in the dip.'

'No. No. She'll be all right. Eh, Kahu?'

Kahu's eyes were shining. 'Oh *yes*. Can I go, Nanny?'

'All right then,' Nanny Flowers grumbled. 'But tomorrow you have to be my mate in the kumara patch. Okay?'

So it was that Kahu became the mascot for me and the boys and it only seemed natural, after a while, for us to take her with us wherever we went — well, most places anyway and only when Nanny Flowers didn't want her in the kumara patch.

But that first night I got my beans from the old lady.

'*Hoi*,' she said, and *bang* came her hand. 'What did you do with Kahu at the shed? She's tuckered out.'

'Nothing,' I squealed. *Biff* came her fist at my stomach. 'She just helped us sheepo and sweep the board and press the wool and pick up the dags and —'

Swish came the broom. 'Yeah,' Nanny Flowers said. 'And I'll bet all you beggars were just lying back and having a good *smoke*.'

You could never win with Nanny Flowers.

At that time the wananga sessions were proving to be very popular. All of us felt the need to understand more about our roots. But Nanny Flowers still grumbled whenever we had our hui. She would sit with Kahu in her arms, rocking in the chair on the verandah, watching the men walk past.

'There go the Ku Klux Klan,' she would say loudly so that we could all hear.

Poor Kahu, she could never keep away from our wananga. She would always try to listen in at the doorway to the meeting house.

'Haere atu koe,' Koro Apirana would thunder. But there was one wananga that Kahu could not eavesdrop on, and that was the one which Koro Apirana led when he took us out in a small flotilla of fishing boats to have a lesson on the sea.

'In our village,' Koro Apirana told us, 'we have always endeavoured to live in harmony with Tangaroa's kingdom and the guardians therein. We have made offerings to the sea god to thank him and when we need his favour, and we have called upon our guardians whenever we are in need of help. We have blessed every new net and new line to Tangaroa. We have tried not to take food with us in our boats when we fish because of the tapu nature of our task.'

The flotilla was heading out to sea.

'Our fishing areas have always been placed under the protective custody of the guardians,' Koro Apirana said. 'In their honour we have often placed talismanic mauri as shrines. In this way the fish have been protected, attracted to the fishing grounds, and thus a plentiful supply has been assured. We try never to overfish for to do so would be to take greedy advantage of Tangaroa and would bring retribution.'

Then we reached the open sea and Koro Apirana motioned that we should stay close to him.

'All of our fishing grounds, banks and rocks have had names assigned to them and the legends surrounding them have been commemorated in story, song or proverb. Where our taunga ika have no local identification, like a reef or upjutting rock, we have taken the *fix* from prominent cliffs or mountains on the shore. Like *there* and *mark*. And *there* and *mark*. In this way the toka hapuku and grounds of other fish have always been known. And we have tried never to trespass on the fishing grounds of others because *their* guardians would recognise us as interlopers. In this respect, should we ever be in unfamiliar sea, we have surrounded ourselves with our own water for protection.'

Then Koro Apirana's voice dropped and, when he resumed his korero, his words were steeped with sadness and regret. 'But we have not always kept our pact with Tangaroa, and in these days of commercialism it is not always easy to resist temptation.

So it was when I was your age. So it is now. There are too many people with snorkelling gear, and too many commercial fishermen with licences. We have to place rahui on our fishing beds, boys, otherwise it will be just like the whales —'

For a moment Koro Apirana hesitated. Far out to sea there was a dull booming sound like a great door opening, a reminder, a memory of something downward plunging. Koro Apirana shaded his eyes from the sun.

'Listen, boys,' he said, and his voice was haunted. 'Whakarongo. Once there were many of our protectors. Now there are few. *Listen how empty our sea has become.*'

In the evening after our lesson on the sea we assembled in the meeting house. The booming on the open waters had heralded the coming of a rainstorm like a ghostly wheke advancing from the horizon. As I went into the meeting house I glanced up at our ancestor, Paikea. He looked like he was lifting his whale through the spearing rain.

Koro Apirana led us in karakia, a prayer to bless the wananga. Then, after the mihimihi, he told us of the times which had brought the silence to the sea.

'I was a boy of seven years' age,' he began, 'when I went to stay with my uncle who was a whaler. I was too young to know any better, and I didn't understand then, as I do now, about our tipua. At that time whaling was one of the great pastimes and once the bell on the lookout had been sounded you'd see all the whaling boats tearing out to sea, chasing after a whale. Doesn't matter what you were doing, you'd drop everything, your plough, your sheep clippers, your schoolbooks, *everything*.

I can still remember seeing everyone climbing the lookout, like white balloons. I followed them and far out to sea I saw a herd of whales.'

The rain fell through his words. 'They were the most beautiful sight I had ever seen.' He made a sweeping gesture. 'Then, down by the slipway, I could see the longboats being launched into the sea. I ran down past the sheds and the pots on the fires were already being stoked to boil down the blubber. All of a sudden my uncle yelled out to me to get on his boat with him. So there I was, heading out to sea.'

I saw a spiky kina sneaking a look through the door. 'That's when I saw the whales really close,' Koro Apirana said. 'There must have been sixty of them at least. I have never forgotten, never. They had mana. They were so powerful. Our longboat got so close to one that I was able to reach out and touch the skin.' His voice was hushed with awe. 'I felt the ripple of power beneath the skin. It felt like silk. Like a god. Then the harpoons began to sing through the air. But I was young, you see, and all I could feel was the thrill, like when you do a haka.'

He paused, mesmerised. 'I can remember that when a whale was harpooned it would fight like hang. Eventually it would spout blood like a fountain, and the sea would be red. Three or four other boats would tow it ashore to the nearest place and cut it up and share out the meat and the oil and everything. When we started to strip the blubber off the whale in the whaling station, all the blood flowed into the channel. Blind eels would come up with the tide to drink the blood.'

I heard Kahu weeping at the doorway. I edged over to her and when she saw me she put her arms around my neck.

'You better go home,' I said, 'before Koro Apirana finds out you're here.'

But she was so frightened. She was making a mewling sound in her throat. She seemed immobilised by terror.

Inside, Koro Apirana was saying, 'Then, when it was all finished we would cut huge slabs of whale meat and sling them across our horses and take them to our homes —'

Suddenly, before I could stop her, Kahu wrenched away from me and ran into the meeting house.

'No, Paka, *no!*' she screamed.

His mouth dropped open. 'Haere atu koe,' he shouted.

'Paka. Paka, no!'

Grimly, Koro Apirana walked up to her, took her by the arms and virtually hurled her out. 'Haere atu. *Haere,*' he repeated. The sea thundered ominously. The rain fell like spears.

Kahu was still crying, three hours later. Nanny Flowers was livid when she heard about what happened.

'You just keep her away from the meeting house,' Koro Apirana said. 'That's all I say. I've told you before. *And* her.'

'My blame,' Kahu wept. 'Love Paka.'

'You *men,*' Nanny Flowers said. 'I can show *you* where you come from.'

'Enough,' Koro Apirana said. He stormed out and that ended the argument.

Later that night Kahu kept sobbing and sobbing. I guess we thought she was still grieving about being growled at, but we know better now. I heard Nanny Flowers going into Kahu's bedroom and comforting her.

'Shift over, moko,' Nanny Flowers soothed. 'Make a little space for your skinny Nanny. There, there.'

'Love Paka.'

'You can *have* him, moko, as soon as I get my divorce tomorrow. There, there.' Nanny was really hurting with love for Kahu. 'Don't you worry, don't you worry. You'll fix him up, the old paka, when you get older, ne.'

In the hiss and roar of the suck of the surf upon the land I listened to Nanny Flowers. After a short while Kahu drifted off to sleep.

'Yes,' Nanny Flowers crooned, 'haere ki te moe. And if you don't fix him,' she whispered, 'then my oath I will.'

Hiss and roar. Ebb and flow.

The next morning I sneaked in to give Kahu a special cuddle, just from me. When I opened the door she was gone. I looked in Koro Apirana and Nanny Flowers' bedroom, but she wasn't there either. Nanny Flowers had pushed Koro onto the floor and had spread herself over the whole bed to make sure he couldn't get back in.

Outside the sea was gentle and serene, as if the storm had never happened. In the clear air I heard a chittering, chattering sound from the beach. I saw Kahu far away, silhouetted on the sand. She was standing facing the sea and listened to voices in the surf. *There, there, moko. There, there.*

Suddenly Kahu turned and saw me. She ran toward me like a seagull. 'Uncle Rawiri!'

I saw three silver shapes leaping into the dawn.

AUTUMN

Season of the Sounding Whale

Nine

If you ask me the name of this house, I shall tell you. It is Te Kani. And the carved figure at the apex? It is Paikea, it is Paikea. Paikea swam, *hei*. The sea god swam, *hei*. The sea monster swam, *hei*. And Paikea, you landed at Ahuahu. You changed into Kahutia Te Rangi, *aue*. You gave your embrace to the daughter of Te Whironui, *aue*, who sat in the stern of the canoe. *Aue*, *aue*, and now you are a carved figurehead, old man.

The sea trench, Hawaiki. The Place of the Gods. The Home of the Ancients. The whale herd hovered in the goldened sea like regal airships. Far above, the surface of the sea was afire with the sun's plunge from day into night. Below lay the sea trench. The herd was waiting for the sign from their ancient leader that it should descend between the protective walls of the trench and flow with the thermal stream away from the island known as the Place of the Gods.

  But their leader was still mourning. Two weeks earlier the herd had been feeding in the Tuamotu Archipelago when suddenly a flash of bright light had scalded the sea and giant tidal soundwaves had exerted so much pressure that internal ear canals had bled. Seven young calves had died. The ancient whale remembered this occurrence happening before; screaming a lament of condemnation, he had led them away in front of the lethal tide that he knew would come. On that pellmell, headlong and mindless escape, he had noticed more cracks in the ocean floor, hairline fractures indicating serious damage below the crust of the earth. Now, some weeks later, the leader was still unsure about the radiation level in the sea trench. He was fearful of the contamination seeping from Moruroa. He was afraid of the genetic effects of the undersea radiation on the remaining herd and calves in this place which had once, ironically, been the womb of the world.

The elderly females tried to nurse his nostalgia, but the ancient whale could not stop the rush of memories. Once this place had been crystalline clear. It had been the place of his childhood and that of his golden master too. Following that first disastrous sounding, they had ridden many times above the trench. His golden master had taught the whale to flex his muscles and sinews so that handholds in the skin would appear, enabling the rider to ascend to the whale's head. There, further muscle contractions would provide saddle and stirrups. And when the whale sounded, he would lock his master's ankles with strong muscles and open a small breathing chamber, just behind his spout. In the space of time, his master needed only to caress his left fin, and the whale would respond.

Suddenly, the sea trench seemed to pulsate and crackle with a lightswarm of luminescence. Sparkling like a galaxy was a net of radioactive death. For the first time in all the years of his leadership, the ancient whale deviated from his usual primeval

*track. The herd ascended to the surface. The decision was made
to seek before time the silent waters of the Antarctic. But the
elderly females pealed their anxieties to one another because the
dangerous islands were also in that vicinity. Nevertheless they
quickly followed their leader away from the poisoned water.
They were right to worry because the ancient whale could only
despair that the place of life, and the Gods, had now become a
place of death. The herd thundered through the sea. Haumi e,
hui e, taiki e.*

Ten

The next year Kahu turned four and I decided it was about time I went out to see the world. Koro Apirana thought it was a good idea but Nanny Flowers didn't like it at all.

'What's wrong with Whangara?' she said. 'You got the whole world right here. Nothing you can get anywhere else that you can't get here. You must be in *trouble*.'

I shook my head. 'No, I'm clean,' I answered.

'Then there must be a girl you're running away from. She looked at me suspiciously, and poked me between the ribs. 'You been up to mischief, eh?'

I denied that too. Laughing, I eased myself up from the chair and did a Clint Eastwood. 'Let's just say, Ma'am,' I drawled, then went for my six-gun, 'that there's not enough room in this here town for the two of us.'

Over the following four months I put in double time at the Works and got my Air New Zealand ticket. The boys took up a collection and gave me a fantastic party. My darling Joyleen Carol cried buckets over me. At the airport I said to Nanny Flowers, 'Don't forget to look after my bike.'

'Don't worry,' she said sarcastically, 'I'll feed it some hay and give it water every day.'

'Give Kahu a kiss from me.'

'Ae,' Nanny Flowers quivered. 'Ma te Atua koe e manaaki. And don't forget to come back, Rawiri, or else —'

She pulled a toy water-pistol from her kete.

'*Bang*,' she said.

I flew to Australia.

Unlike Kahu, my pito couldn't have been put in the ground at Whangara because I didn't return there until four years later. I discovered that everything I'd been told about Aussie was true: it was big, bold, brassy, bawdy and beautiful. When I first arrived I stayed in Sydney with my cousin, Kingi, who had an apartment in Bondi. I hadn't realised that there were so many other Maoris over there (I thought I'd be the first) and after a while I realised why it was nicknamed 'Kiwi Valley'. Wherever you went, the pubs, the shows, the clubs, the restaurants, the movies, the theatres, you could always count on bumping into a cousin. In some hotels, above the noise and buzz of the patrons, you were bound to hear somebody shouting to somebody else, 'Kia ora, cous!'

I was like a kid in a great big toyshop, wanting to touch everything. Whangara wasn't as big as *this*, with its teeming city streets, glass skyscrapers, glitter and glitz. Nor could Friday night in the town ever compare with the action in the Cross, that part of Sydney to which people thronged, either to look or be looked at. People were selling anything and everything up the Cross and if you wanted to buy you just 'paid the man'.

It was there that I came upon my cous Henare, who was now wearing a dress, and another cous, Reremoana, who had changed her name to Lola L'Amour and had red hair and fishnet stockings. I couldn't understand Kingi's attitude at all; he was always trying to cross the street whenever he saw a cous he didn't want to be seen with. But I would just bowl along regardless and yell, 'Kia ora cous!'

As far as I could see, they were living the way they wanted to and no matter what changes they had made to themselves or their lives, a cous was a cous. I guess also that I didn't feel that much different: I looked much the same as they did, with my leather jacket and pants matching their own gear with its buckles and scarves and whips. 'What game are you into?' they would tease. 'What game?' They would joss and kid and joke around and sometimes we would meet up later at some party or other. But always, in the early morning, when the sunlight was beginning to crack the midnight glamour, the memories would come seeping through. 'How's our Nanny? How's our Koro? If you write to them, don't tell them that you saw us like *this*.'

In the search for fame, fortune, power and success, some of my cousins had opted for the base metal and not the gold. They may have turned their lives upside down in the process, like Sydney Bridge's reflection in the harbour, but they always craved the respect of our whanau. They weren't embarrassed, but hiding the way they lived was one way of maintaining the respect. There was no better cloak than those starry nights under the turning Southern Cross.

Kingi and I got along fine, but when I found a mate of my own, I moved in with him. I had gotten a job working as a brickie and had also started playing League. It was through League that I met my buddy, Jeff, who told me he was looking for someone to share his flat. Jeff was a friendly, out-front guy, quick to laugh, quick to believe and quick to trust. He told me of his family in Mount Hagen, Papua New Guinea, and I told him about mine in Whangara. I also told him about Kahu.

'You'd love her,' I said. 'She's a fantastic looker. Big brown eyes, wonderful figure and lips just waiting to be kissed.'

'Yeah? Yeah?' he asked eagerly.

'And I can tell she'd go for *you*,' I said. 'She's warm and cuddly, great to be with, and she just loves snuggling up close. And —'

Poor Jeff, he didn't realise I was having him on. And as the weeks went by I embellished the story even more. I just couldn't help it. But that's how our friendship was; we were always kidding around or kidding each other.

I must have been in Sydney over a year when the phone call came from my brother Porourangi. Sometimes life has a habit of flooding over you and rushing you along in its overwhelming tide. Living in Aussie was like that: there was always something going on, day and night. If Jeff and I weren't playing League we'd be out surfing (the beach at Whangara was better) or partying with buddies, or hiking out to the Blue Mountains. You could say I had begun drowning in it all, giving myself up to what Kingi would have called 'the hedonistic life of the lotus eater'. Kingi was always one for the big words. He used to tell me that his favourite image of Australia was of Joan Sutherland singing 'Advance Australia Fair', a can of Fosters in one hand, and surfing supremely into Sydney Harbour like an antipodean Statue of Liberty. See what I mean? All those big words? That's Kingi, for sure.

I was still in bed when the telephone rang, so Jeff answered. Next minute, a pillow came flying at me and Jeff yanked me out of bed saying, 'Phone, Rawiri. And I'll talk to *you* later.'

Well, the *good* news was that Porourangi was getting married to Ana. Nanny Flowers had been pestering both of them about it. 'And you know what she's like,' Porourangi laughed. 'Don't bother to come home though,' he said, 'because the wedding is just going to be very small.' Kahu would be the flower girl.

'How is she?' I asked.

'She's five and started school now,' Porourangi said. 'She's still living with Rehua's folks. She missed you very much last summer.'

'Give her a kiss from me,' I said. 'And also kiss our Nanny. Tell everybody I love them. How's Koro?'

'In Nanny's bad books as usual,' Porourangi laughed. 'The sooner they get a divorce the better.'

I wished Porourangi and Ana the very best with their life together. The season of bereavement had been long over for Porourangi and it was time for renewal. Then just before he hung up, he said, 'Oh, by the way, your mate was very inter-ested in Kahu, so I told him she was doing well with her spelling.'

Uh oh. That was the *bad* news. No sooner had I put the phone down than Jeff was onto me.

'Warm and cuddly, huh?'

'No, wait Jeff, I can explain —'

'Big brown eyes and fantastic figure, huh?'

'Jeff, no —' In his hands he had a soggy apple pie.

'Lips just waiting to be kissed?' His eyes gleamed with vengeance.

I should count myself lucky that I had cooked dinner the night before. Had it been Jeff, that apple pie wouldn't have been so scrumptious.

Not long after that Jeff also got a phone call, but the news wasn't so good. His mother called from Papua New Guinea to ask him to come home.

'Your father's too proud to ring himself,' she said, 'but he's getting on, Jeff, and he needs you to help him run the coffee plantation. He's had a run of rotten luck with the workers this year and you know what the natives are like, always drinking.'

'I'll have to go,' said Jeff. I knew he was reluctant to do so. Indeed, one of the reasons why he had come to Sydney was that it was as far from his family as he could get. He loved them deeply, but sometimes love becomes a power game between the ambitions that parents have for their children and the ambitions that children have for themselves. 'But it looks like all my chickens are coming home to roost,' Jeff said ruefully.

'Family is family,' I said.

'Say,' he interrupted. 'You wouldn't like to come with me?'

I hesitated. Ever since speaking to Porourangi I had actually been thinking of going back to Aotearoa. Instead, I said, 'Sure,

I've been a cowboy all my life. Let's saddle up, partner.' So we started to pack up ready to move on out. I rang Whangara to tell Nanny Flowers.

'You're going *where*?' she yelled. As usual she was holding the phone at arm's length.

'To Papua New Guinea.'

'E hika,' she said. 'You'll get eaten up by all them cannibals. What's at Papua New Guinea' — I mouthed the words along with her — 'that you can't get in Whangara? You should come home instead of gallivanting all over the world.'

'I'll be home next summer. I promise.' There was silence at the other end. 'Hullo?'

Koro Apirana came to the telephone. 'E Rawiri, aue,' he said loudly. 'Kei te tangi to koka ki a koe.' There was a tussle at the other end and Nanny Flowers returned.

'I can speak for myself,' she said in a huff. Then, in a soft voice, full of longing, she added, 'All right, e tama. You go to Papua New Guinea. But don't make promises about next summer. Otherwise I will be watching the road, and going down to the bus every day to see if you are on it.'

Tears began to mist my vision. I could just imagine my Nanny walking down the road in summer, Kahu skipping beside her, and sitting on the verge watching the cars going past, and asking the bus driver —

'We love you,' Nanny said.

Waiting and waiting. Then the phone clicked on the handset and she was gone.

Eleven

I was two years with Jeff in Papua New Guinea and while they were productive years, they were not always happy. Jeff's father couldn't come down to Port Moresby to meet us but his mother, Clara, did. Although Jeff had told her I was a Maori it was obvious that I was still too dark. As soon as I stepped off the plane I could almost hear her wondering, 'Oh, my goodness, how am I going to explain this to the women at the Bridge Club?' But she was polite and gracious and kept up a lively chatter on the plane to Mount Hagen.

Tom, Jeff's father, was another story, and I liked him from the start. He was a self-made man whose confidence had not been shattered by his long and debilitating illness. But it was clear that he needed his son to help him. I can still remember the first time I saw Tom. He was standing on the verandah of the homestead, resting his weight on two callipers. He wasn't embarrassed by his disability and when Jeff went up to greet him he simply said, 'Gidday young fella. Glad to have you home.'

Tom had contracted Parkinson's disease. It wasn't until weeks later that I discovered the disease had not only struck at his limbs but also had rendered him partially blind.

The situation was clear. Jeff would have to act as an extension to his father, his arms and legs and eyes. Deskbound, Tom would run the plantation from the homestead and Jeff would translate the instructions into action. As for me, I've always been pretty good at hard work, so it was simply a matter of spitting on my hands and getting down to business.

Putting the plantation back on its feet was a challenge which the countryside really threw at us; I have never known a country which has fought back as hard as Papua New Guinea. I doubt if it can ever be tamed of its temperatures, soaring into sweat zones, for its terrain, so much a crucible of crusted plateaus and valleys, and its tribalism. But we tried, and I think we won some respite from the land, even if only for a short time. Man might carve his moko on the earth but, once he ceases to be vigilant, Nature will take back what man had once achieved to please his vanity.

Sometimes, when you yourself are living life to the full, you forget that life elsewhere also continues to change like a chameleon. For instance, I used to marvel at the nationalism sweeping Papua New Guinea and the attempts by the Government to transplant national identity and customs onto the colonial face of the land. They were doing so despite an amazing set of difficulties: first, Papua New Guinea was fractionalised into hundreds of iwi groups and their reo was spoken in a thousand different tongues; second, there were so many outside influences on Papua New Guinea's inheritance, including their neighbours across the border in Irian Jaya;

and, third, the new technology demanded that the people literally had to live 'one thousand years in one lifetime', from loincloth to the three-piece suit and computer knowledge in a simple step.

In many respects the parallels with the Maori in New Zealand were very close, except that we didn't have to advance as many years in one lifetime. However, our journey was possibly more difficult because it had to be undertaken within Pakeha terms of acceptability. We were a minority and much of our progress was dependent on Pakeha goodwill. And there was no doubt that in New Zealand, just as in Papua New Guinea, our nationalism was also galvanising the people to become one Maori nation.

So it was that in Australia and Papua New Guinea I grew into an understanding of myself as a Maori and, I guess, was being prepared for my date with destiny. Whether it had anything to do with Kahu's destiny, I don't know, but just as I was maturing in my own understanding she too was moving closer and closer to that point where she was in the right place, at the right time, with the right understanding to accomplish the task which had been assigned to her. In this respect there is no doubt in my mind that she had always been the *right* person.

My brother Porourangi has always been a good letter writer and he kept me in touch with the affairs of the people at home. I could tell that his ihi was growing, his spirit, and I appreciated the chiefly kindliness he felt in wanting me to know that although I was far from the family I was not forgotten. Apparently Koro Apirana had now begun a second series of wananga for the young people of the Coast. Our Koro had accepted that Porourangi would be 'the one' in our generation to carry on the leadership of the people, but he was still looking for 'the one' in the present generation. 'He wants to find a young boy,' Porourangi jested, 'to pull the sword out of the stone, someone who has been marked by the Gods for the task. Nobody has so far been able to satisfy him.' Then, in one of his letters, Porourangi made my heart leap with joy. Ana had told him it was about time that Kahu came back to stay in Whangara, with her and Porourangi.

Kahu was then six years old; Rehua's mother had agreed and so Kahu returned. 'Well,' Porourangi wrote, 'you should have seen us all having a tangi at the bus stop. Kahu got off the bus and she has grown so much, you wouldn't recognise her. Her first question, after all the hugging, was 'Where's Paka? Is Paka here?' Nanny Flowers said he was fishing, so she waited and waited all day down at the beach for him. When he came in, she leapt into his arms. But you know our Koro, as gruff as usual. Still, it is really good to have Kahu home.'

In his later letters Porourangi wrote about the problems he felt were facing the Maori people. He had gone with Koro Apirana to Raukawa country and had been very impressed with the way in which Raukawa was organising its youth resources to be in a position to help the people in the century beginning with the year 2000. 'Will *we* be ready?' he asked. 'Will we have prepared the people to cope with the new challenges and the new technology? And will they still be Maori?' I could tell that the last question was weighing heavily on his mind. In this respect we both recognised that the answer lay in Koro Apirana's persistence with the wananga sessions, for he was one of the very few who could pass on the knowledge, the sacred kumara, to us. Our Koro was like an old whale stranded in an alien present, but that was how it was supposed to be because he also had his role in the pattern of things, in the tides of the future.

Near the middle of our second year in Papua New Guinea Jeff and I could afford to relax a little. We took trips to Manus Island and it was there that Jeff put into words the thoughts that had been on my mind for some months.

'You're getting homesick, aren't you Rawiri?' he said.

We had been diving in the lagoon, and in that wondrous blue water, I had picked up a shining silver shell from the reef. I had taken it back to the beach and was listening to the sea whispering to me from the shell's silver whorls.

'A little,' I replied. Many things were coming to a head for me on the plantation, and I wanted to avoid a collision. Jeff and I were getting along okay but his parents were pushing him ever so gently in the right direction, to consort with his own kind in the clubs and all the parties of the aggressively expatriate. On my part, this had thrown me more into the company of the 'natives', like Bernard who had more degrees than Clara had chins, and Joshua, who both worked on the farm. In so doing I had broken a cardinal rule and my punishment was ostracism.

'We've come a long way together,' Jeff said.

'We sure have,' I laughed. 'And there's still a way to go yet.'

Then Jeff said, 'I want to thank you. For everything. But if you have to go, I'll understand.'

I smiled at him, reflectively. I placed the shell back to my ear. *Hoki mai, hoki mai ki te wa kainga*, the sea whispered.

Jeff and I returned to the plantation the next day. There was a letter waiting from Porourangi. Ana was expecting a baby, and the whole family were hoping that the child would be a son. 'Of all of us,' Porourangi wrote, 'Kahu seems to be the most excited. Koro Apirana, too, is over the moon.'

The letter had the effect of making me realise how much time had passed since I had been in the company of my whanau, and I felt a sudden keenness, like pincers squeezing my heart, to hold them all in my arms. *Hoki mai, hoki mai.*

Then three events occurred which convinced me that I should be homeward bound. The first happened when Jeff and his parents were invited to a reception hosted in Port Moresby for a young expatriate couple who'd just been wed. At first Clara's assumption was that I would stay back and look after the plantation, but Jeff said I was 'one of the family' and insisted I accompany them. Clara made it perfectly obvious that she was embarrassed by my presence and I was very saddened, at the reception, to hear her say to another guest, 'He's a friend of Jeff's. You know our Jeff, always bringing home dogs and strays. But at least he's not a native.' Her laughter glittered like knives.

But that was only harbinger to the tragedy which took place when we returned to Mount Hagen. We had parked the station wagon at the airport and were driving home to the plantation.

Jeff was at the wheel. We were all of us in a merry mood. The road was silver with moonlight. Suddenly, in front of us, I saw a man walking along the verge. I thought Jeff had seen him too and would move over to the middle of the road to pass him. But Jeff kept the station wagon pointed straight ahead.

The man turned. His arms came up, as if he was trying to defend himself. The front bumper crunched into his thighs and legs and he was catapulted into the windscreen which smashed into a thousand fragments. Jeff braked. The glass was suddenly splashed with blood. I saw a body being thrown ten metres to smash on the road. In the headlights and steam, the body moved. Clara screamed. Tom said, 'Oh my God.'

I went to get out. Clara screamed again, 'Oh no. No. His tribe could be on us any second. Payback, it could be payback for us. It's only a native.'

I pushed her away. Tom yelled, 'For God's sake, Rawiri, try to understand. You've heard the stories —'

I couldn't comprehend their fear. I looked at Jeff but he was just sitting there, stunned, staring at that broken body moving fitfully in the headlights. Then, suddenly Jeff began to whimper. He started the motor.

'Let me out,' I hissed. 'Let me *out*. That's no native out there. That's *Bernard*.' A cous is a cous.

I yanked the door open. Clara yelled out to Jeff, 'Oh, I can see them.' Shadows on the road. 'Leave him here. Leave him.' Her words were high-pitched, frenzied. 'Oh. Oh. Oh.'

The station wagon careered past me. I will never forget Jeff's white face, so pallid, so fearful.

The second event occurred after the inquest. Bernard had

74

died on the road that night. Who's to say that he would have lived had we taken him to hospital?

It was an accident, of course. A native walking carelessly on the side of the road. A cloud covering the moon for a moment. The native shouldn't have been there anyway. It could have happened to anybody.

'I don't blame you,' I said to Jeff. 'You can't help being who you are.' But all I could think of was the waste of a young man who had come one thousand years to his death on a moonlit road, the manner in which the earth must be mourning for one of its hopes and its sons in the new world, and the sadness that a friend I thought I had would so automatically react to the assumptions of his culture. *And would I be next?* There was nothing further to keep me here.

It was then that another letter came from Porourangi. The child, a girl, had been born. Naturally, Koro Apirana was disappointed and had blamed Nanny Flowers again. In the same envelope was another letter, this one from Kahu.

'Dear Uncle Rawiri, kia ora. How are you? We are well at Whangara. I have a baby sister. I like her very much. I am seven. Guess what, I am in the front row of our Maori culture group at school. I can do the poi. We are all lonesome for you. Don't forget me, will you. Arohanui. Kahutia Te Rangi.'

Right at the bottom of Kahu's letter Nanny Flowers had added just one word to express her irritation with my long absence from Whangara. *Bang.*

I flew out of Mount Hagen the following month. Jeff and I had a fond farewell, but already I could feel the strain between us. Clara was as polite and scintillating as usual. Tom was bluff and hearty.

'Goodbye, fella,' Tom said. 'You're always welcome.'

'Yes,' Jeff said. 'Always.' *Each to his own.*

The plane lifted into the air. Buffetted by the winds it finally stabilised and speared through the clouds.

Ah yes, the clouds. The third event had been a strange cloud formation I had seen a month before above the mountains. The clouds looked like a surging sea and through them from far away a dark shape was approaching, slowly plunging. As it came closer and closer I saw that it was a giant whale. On its head was a sacred sign, a gleaming moko. Haumi e, hui e, *taiki e.*

These photos come from *Whale Rider*, the movie adaptation
of the novel directed by Niki Caro.

The coastal settlement of Whangara, Aotearoa/New Zealand.

Nanny Flowers (Vicky Haughton) and Koro Apirana (Rawiri Paratene)
take baby Pai/Kahu into their arms and into their home.

Pai (Keisha Castle-Hughes) grows into an energetic, charismatic young woman.

The family at home; (from left to right) Nanny Flowers, Uncle Rawiri (Grant Roa), who narrates the novel, Paikea, Shilo (Rachel House) and Porourangi (Cliff Curtis).

Pai and her father Porourangi are very close, yet he choses to follow his career
away from his daughter.

Koro and Pai share an unshakable bond, but Koro believes a male heir must receive the knowledge of his tipuna (ancestors).

Pai proudly recites her whakapapa (genealogy) in a community celebration. She searches for Koro, who remains noticeably absent.

Koro convenes a wananga (school) to instruct local boys in the ancient customs of his people, but Pai is excluded because she is a girl.

Pai remains optimistic and determined to learn what the boys are learning.

Nanny and Koro join the embattled community of Whangara when whales become stranded on the shore. Despite the efforts of the people, the stranded whales begin to weaken.

Pai is drawn to the whales, and intuitively begins to communicate with the stranded herd leader.

Pai climbs atop the great whale and leads the herd to the safety of the ocean depths. She is rescued from the sea but the journey has taken a toll.

Porourangi and Rawiri launch the tribe's waka (canoe), symbolising the renewal of mauri (life force) that Pai has sparked.

Twelve

I wish I could say that I had a rapturous return. Instead, Nanny Flowers growled at me for taking so long getting home, saying, 'I don't know why you wanted to go away in the first place. After all —'

'I know, Nanny,' I said. 'There's nothing out there that I can't get here in Whangara.'

Bang came her hand. 'Don't *you* make fun of me too,' she said, and she glared at Koro Apirana.

'Huh?' Koro said. 'I didn't say nothing.'

'But I can hear you *thinking*,' Nanny Flowers said, 'and I know when you're funning me, you old paka.'

'Te mea te mea,' Koro Apirana said. 'Te mea te mea.'

Before Nanny Flowers could explode I gathered all of her in my arms, and there was much more of her now than there had been before, and kissed her. 'Well,' I said, 'I don't care if you're not glad to see me, because I'm glad to see *you*.'

Then I handed her the present I had bought her on my stopover in Sydney. You would have thought she'd be pleased but instead, *smack* came her hand again.

'You think you're smart, don't you,' she said.

I couldn't help it, but I had to laugh. 'Well how was I to

know you'd put on weight!' My present had been a beautiful dress which was now three sizes too small.

That afternoon I was looking out the window when I saw Kahu running along the road. School had just finished.

I went to the verandah to watch her arrival. Was this the same little girl whose pito had been put in the earth those many years ago? Had seven years really gone past so quickly? I felt a lump at my throat. Then she saw me.

'Uncle Rawiri!' she cried 'You're back!'

The little baby had turned into a doe-eyed, long-legged beauty with a sparkle and infectious giggle in her voice. Her hair was unruly, like an Afro, but she had tamed it into two plaits today. She was wearing a white dress and sandals. She ran up the steps and put her arms around my neck.

'Kia ora,' she breathed as she gave me a hongi.

I held her tightly and closed my eyes. I hadn't realised how much I had missed the kid. Then Nanny Flowers came out and said to Kahu, 'Enough of the loving. You and me are working girls! Haere mai! Kia tere!'

'Nanny and me are hoeing the kumara patch,' Kahu smiled. 'I come every Wednesday to be her mate, when she wants a rest from Koro.' Then she gave a little gasp and took my hand and pulled me around to the shed at the back of the house.

'Don't be too long, Kahu,' Nanny Flowers shouted. 'Those kumara won't wait all day.'

Kahu waved okay. As I followed her I marvelled at the stream of conversation which poured out of her. 'I've got a baby sister now, uncle, she's a darling. Her name is Putiputi after Nanny Flowers. Did you know I was top of my class this year? And I'm the leader of the culture group too. I love singing the Maori songs. Will you teach me how to play the guitar? Oh, *neat*. And Daddy and Ana are coming to see you tonight once Daddy gets back from work. You bought me a *present? Me?* Oh where is it, where is it! You can show me later, ay. But I want you to see this first —'

She opened the door to the shed. Inside I saw a gleam of shining silver chrome. Kahu put her arms around me and kissed me again. It was my motorbike.

'Nanny Flowers and I have been cleaning it every week,' she said. 'She used to cry sometimes, you know, when she was cleaning it. Then she'd get scared that she might cause some rust.'

I just couldn't help it; I felt a rush of tears to my eyes. Concerned, Kahu stroked my face.

'Don't cry,' she said. 'Don't cry. It's all right, Uncle Rawiri. There, there. You're home now.'

Later that night Porourangi arrived. Among the family he was the one who seemed to have aged the most. He proudly showed me the new baby, Putiputi.

'Another girl,' Koro Apirana said audibly, but Porourangi took no notice of him. We were used to Koro's growly ways.

'Turituri to waha,' Nanny Flowers said. 'Girls can do anything these days. Haven't you heard you're not allowed to discriminate against women any more? They should put you in the jailhouse.'

'I don't give a hang about women,' Koro Apirana said. 'You still haven't got the mana.'

It was then that Nanny Flowers surprised us all. 'Oh, te mea te mea, you old goat,' she said.

We had a big family kai that night with Maori bread and crayfish and lots of wai to drink. Nanny had invited the boys over and they arrived with a roar and a rush of blue smoke and petrol fumes. It was almost as if I had never left. The guitars came out and the voices rang free to make the stars dance with joy. Nanny Flowers was in her element, playing centre stage to her whanau, and one of the boys got her up to do a hula.

'Anei,' he cried with delight. 'Te Kuia o te whanau o Whangara!'

There was a roar of laughter at that one, and Kahu came running up to me, saying, 'See how we love you, uncle? We killed the fatted calf for you, just like the Bible says.' She hugged me close and then skipped away like a songbird.

Then Porourangi was there. 'Is it good to be home?' he asked cautiously.

'Yes,' I breathed. 'Just *fantastic*. How has it been?'

'Much the same as ever,' Porourangi said. 'And you know our Koro. He's still looking.'

'What for?'

'The one who can pull the sword,' Porourangi laughed hollowly. 'There are a few more young boys he's found. One of them may be the one.'

Porourangi fell silent. I saw Koro Apirana rocking in his chair, back and forth, back and forth. Kahu came up to him and put her hand in his. He pushed her away and she dissolved into the dark. The guitars played on.

Over the following weeks it was clear to me that Koro Apirana's search for 'the one' had become an obsession. Ever since the birth of Kahu's young sister he had become more intense and brooding. Perhaps aware of his own mortality, he wanted to make sure that the succession in the present generation was done, and done well. But in doing so he was pushing away the one who had always adored him, Kahu herself.

'You'd think the sun shone out of his —' Nanny Flowers said rudely. Kahu had come to the homestead that morning riding a horse, with the news that she'd come first in her Maori class. Nanny Flowers had watched as Koro Apirana had dismissed the young girl. 'I don't know why she keeps on with him.'

'I know why,' I said to Nanny Flowers. 'You remember when she bit his toe? Even then she was telling him, "Yeah, don't think you're going to keep me out of this!" '

Nanny Flowers shrugged her shoulders. 'Well, whatever it is, Kahu is sure a sucker for punishment, the poor kid. Must be my Muriwai breed. Or Mihi.'

Mihi Kotukutuku had been the mother of Ta Eruera, who had been Nanny's cousin, and we loved the stories of Mihi's exploits. She was a big chief, descended as she was from Apanui, after whom Nanny's tribe was named. The story we liked best was the one telling how Mihi had stood on a marae at Rotorua. 'Sit down,' a chief had yelled, enraged. 'Sit down,' because women weren't supposed to stand up and speak on the marae. But Mihi had replied, 'No *you* sit down! I am a senior line to yours!' Not only that, but Mihi had then turned her back to him, bent over, lifted up her petticoats and said, 'Anyway, here is the place where you come from!' In this way Mihi had emphasised that all men are born of women.

We sat there on the verandah, talking about Kahu and how beautiful she was, both inside and outside. She had no guile. She had no envy. She had no jealousy. As we were talking, we saw Koro Apirana going down to the marae where seven boys were waiting.

'Them's the contenders,' Nanny Flowers said. 'One of them's going to be the Rocky of Whangara.'

Suddenly Kahu arrived, dawdling from the opposite direction. She looked so disconsolate and sad. Then she saw Koro Apirana. Her face lit up and she ran to him, crying 'Paka! Oh! Paka!'

He turned to her quickly. 'Haere atu koe,' he said. 'Haere atu. You are of no use to me.'

Kahu stopped in her tracks. I thought she would cry, but she knitted her eyebrows and gave him a look of such frustration that I could almost hear her saying to herself, 'You just wait, Paka, you just wait.' Then she skipped over to us as if nothing had happened.

I was lucky enough to get a job in town stacking timber in a timber yard and delivering orders to contractors on site. Every morning I'd beep the horn of my motorbike as I passed Porourangi's, to remind Kahu it was time for her to get out of bed for school. I soon began to stop and wait until I saw her head poking above the window-sill to let me know she was awake. 'Thank you, Uncle Rawiri,' she would call as I roared off to work.

Sometimes after work I would find Kahu waiting at the highway for me. 'I came down to welcome you home,' she would explain. 'Nanny doesn't want any help today. Can I have a ride on your bike? I *can*? Oh, *neat*!' She would clamber on behind me and hold on tight. As we negotiated the track to the village I would be swept away by her ingenuous chatter. 'Did you have a good day, uncle? I had a *neat* day except for maths, yuk, but if I want to go to university I have to learn things I don't like. Did you go to university, uncle? Koro says it's a waste of time for a girl to go. Sometimes I wish I wasn't a girl. Then Koro would love me more than he does. But I don't mind. What's it like being a boy, uncle? Have you got a girlfriend? There's a boy at school who keeps following me around. I said to him that he should try Linda. She likes boys. As for me, I've only got one boyfriend. No, *two*. No, *three*. Koro, Daddy and *you*. Did you miss me in Australia, uncle? Did you like Papua New Guinea? Nanny Flowers thought you'd end up in a pot over a fire. She's a hardcase, isn't she! You didn't forget me, uncle, did you? You didn't, eh? Well, thank you for the ride, Uncle Rawiri. See you tomorrow. Ka kite ano.' With an ill-aimed kiss and a hug, and a whirl of white dress, she would be gone.

The end of the school year came, and the school break-up ceremony was to be held on a Friday evening. Kahu had sent invitations to the whole family and included the boys in the list. 'You are cordially invited,' the card read, 'to the school prizegiving and I do hope you are able to attend. No RSVP is required. Arohanui, Kahutia Te Rangi. P.S. No leather jackets please, as this is a formal occasion. P.P.S. Please park all motorbikes in the area provided and not in the Headmaster's parking space like last year. I do not wish to be embraced again.'

On the night of the break-up ceremony, Nanny Flowers said to me, as she was getting dressed. 'What's this word "embraced"?'

'I think she means "embarrassed",' I said.

'Well, how do I look?' Nanny asked.

She was feeling very pleased with herself. She had let out the dress I had bought her and added lime-green panels to the sides. Nanny was colour blind and thought they were red. I gulped hard. 'You look like a duchess,' I lied.

'Not like a queen?' Nanny asked, offended. 'E hara, I'll soon fix that.' Oh no, not the *hat*. It must have looked wonderful in the 1930s but that was ages ago. Ever since, she had added a bit of this and a bit of that until it looked just like her kumara patch.

'Oh,' I swallowed, 'you look out of this world.'

She giggled coyly. We made our way out to Porourangi's car. Kahu's face gleamed out at us.

'Oh you look lovely,' she said to Nanny, 'but there's something wrong with your hat.' She made a space for Nanny and said to her, 'Come and sit by me, darling, and I'll fix it for you.'

Porourangi whispered to me, 'Couldn't you stop the old lady? Her and her blinking hat.'

I was having hysterics. In the back seat Kahu was adding some feathers and flowers and what looked like weeds. The strange thing was that in fact the additions made the hat just right.

The school hall was crowded. Kahu took us to our places and sat us down. There was an empty seat beside Nanny with 'Reserved' on it.

'That's for Koro when he comes,' she said. 'And don't the boys look *neat*?' At the back of the hall the boys were trying to hide behind their suit jackets.

Nanny Flowers jabbed Porourangi in the ribs. 'Didn't you tell that kid?' she asked.

'I didn't have the heart,' he whispered.

For the rest of that evening the seat beside Nanny Flowers remained empty, like a gap in a row of teeth. Kahu seemed to be in everything: the school choir, the skits and the gymnastics, and after every item she would skip back to us and say, 'Isn't Koro here yet? He's missing the best part.'

Then the second half of the programme began. There was Kahu in her piupiu and bodice, standing so proudly in front of the school cultural group. 'Hope!' she yelled. 'Tena i whiua!' she ordered. And as she sang, she smiled a brilliant smile at all of us. Her voice rang out with pride.

'That young girl's a cracker,' I overheard someone say. But my heart was aching for her and I wanted to leave. Nanny Flowers gripped me hard and said, 'No, we all have to sit here, like it or not.' Her lips were quivering.

The action songs continued, one after another, and I could see that Kahu had realised that Koro Apirana was not going to arrive. The light kept dimming, gradually fading from her face, like a light bulb flickering. By the time the bracket was concluded she was staring down at the floor trying not to see us. She looked as if she was feeling ashamed, and I loved her all the more for her vulnerability.

We tried to bolster her courage by clapping loudly, and we were rewarded by a tremulous smile playing on her face. It was then that the headmaster stepped forward. He made an announcement: one of the students would read the speech which had won the East Coast primary schools contest. What was remarkable, he said, was that the student had given it entirely in her own tongue, the Maori language. He called for Kahutia Te Rangi to come forward.

'Did you know about this?' Nanny Flowers asked.

'No,' said Porourangi. 'Come to think of it, she did mention she had a surprise. For her Koro —'

To the cheers of her schoolmates Kahu advanced to the front of the stage.

'E nga rangatira,' Kahu began, 'e nga iwi,' she looked at Koro Apirana's empty seat, 'tena koutou, tena koutou, tena koutou katoa.' There were stars in her eyes, like sparkling tears. 'Tenei korero he korero aroha mo taku koroua, ko Apirana.'

Nanny Flowers gave a sob, and tears began to flow down her cheeks.

Kahu's voice was clear and warm as she told of her love for her grandfather and her respect for him. Her tones rang with pride as she recited his whakapapa and ours. She conveyed how grateful she was to live in Whangara and that her main aim in life was to fulfil the wishes of her koro and of the iwi.

And I felt so proud of her, so proud, and so sad that Koro Apirana was not there to hear how much she loved him. And

I wanted to shout, Kia kaha, Kia manawanui, to this young girl who was not really so brave and who would have liked the support of the one person who was never there — her Koro, Apirana. At the end of the speech I leapt to my feet to haka my support for her. Then the boys were joining in, and Nanny Flowers was kicking off her shoes. 'Uia mai koia, whakahuatia ake ko wai te whare nei e? Ko Te Kani —' The sadness and the joy swept us all away in acknowledging our tamahine, but we knew that her heart was aching for Koro Apirana.

In the car, later, Porourangi said, 'Your Koro couldn't make it tonight, darling.'

'That's all right, Daddy. I don't mind.'

Nanny Flowers hugged her fiercely. 'I tell you, Kahu, tomorrow I'm really getting a divorce. Your Koro can go his way and I'll go mine.'

Kahu put her face against Nanny Flowers' cheeks. Her voice was drained and defeated. 'It's not Paka's fault, Nanny,' she said, 'that I'm a girl.'

Thirteen

Two weeks after the school break-up ceremony, Koro Apirana took the young boys from the wananga onto the sea. It was early morning as he put them in his boat and headed out past the bay where the water suddenly turned dark green.

When the sun tipped the sea, Koro Apirana began a karakia. He had a carved stone in his hand and suddenly he threw it into the ocean. The boys watched until they could see it no longer.

'One of you must bring that stone back to me,' Koro Apirana said. 'Go now.'

The boys were eager to prove themselves but the stone had gone too deep. Some were afraid of the darkness. Others were unable to dive so far down. Despite valiant attempts they could not do it.

Koro Apirana's face sagged. 'Okay, boys, you've done well. Let's get you all home.'

When he got back to the homestead, Koro Apirana shut himself in the bedroom. Slowly, he began to tangi.

'What's wrong with my Koro?' Kahu asked. She was sitting with me on the verandah. 'Is it because of the stone?'

'How did you know about that?' I asked, astonished.

'One of the boys told me,' Kahu said. 'I wish I could make Paka happy again.' Her eyes held a hint of gravity.

The next morning I was up early, intending to go out onto the sea in my dinghy. To my surprise, Kahu was waiting at the door in her white dress and sandals. There were white ribbons in her pigtails.

'Can I come for a ride, Uncle Rawiri?' she asked.

I couldn't really say no, so I nodded my head. Just as we were ready to leave, Nanny Flowers yelled out, '*Hoi*, wait for me!' She had decided to join us. 'I can't stand to hear the old paka feeling sorry for himself. Mmmm, what a beautiful day! Kei te whiti te ra.'

We rowed out past the bay and Kahu asked again about the stone.

'What stone!' Nanny Flowers said.

So I told her, and Nanny wanted to be shown where it had been dropped into the water. We went out into the ocean where it suddenly turned indigo.

'E hika,' Nanny said. 'No wonder those boys couldn't get it. This is *deep*.'

'Does Koro Apirana really want it back?' Kahu asked.

'Yeah, I suppose he must,' Nanny Flowers said, 'the old paka. Well, serve him right for —'

Kahu said simply, 'I'll get it.'

Before we could stop her she stood up and dived overboard. Until that moment I had never even known she could swim.

Nanny's mouth made a big O. Then the breath rushed into her lungs and she screamed, 'Aue, moko!' She jabbed me hard and said, 'Go after her, Rawiri. *Go*.' She virtually pushed me over the side of the rowboat.

'Give me the diving mask,' I yelled. Nanny Flowers threw it at me and quickly I put it on. I took three deep breaths and did a duck dive.

I couldn't see her. The sea looked empty. There was only a small stingray flapping down towards the reef.

Then I got a big fright because the stingray turned around and, smiling, waved at me. It was Kahu in her white dress and sandals, dog paddling down to the sea floor, her braids floating around her head.

I gasped and swallowed sea water. I came to the surface coughing and spluttering.

'Where is she!' Nanny Flowers screamed. 'Has she drowned? Aue, taku mokopuna.' And before I could stop her she jumped in beside me, just about emptying the whole ocean. She didn't even give me a chance to explain as she grabbed the mask off me and put it on. Then she tried to swim underwater, but her dress was so filled with air that no matter how hard she tried she remained on the surface like a balloon with legs kicking out of it. I doubt if she could have gotten deeper anyway because she was so fat she couldn't sink.

'Oh my mokopuna,' Nanny Flowers cried again. But this time I told her to take a deep breath and, when she was looking underwater, to watch where I would point.

We went beneath the surface. Suddenly I pointed down. Kahu was searching the reef, drifting around the coral. Nanny Flowers' eyes widened with disbelief.

Whatever it was Kahu was searching for, she was having difficulty finding it. But just then white shapes came speeding out of the dark towards her. I thought they were sharks, and Nanny Flowers began to blow bubbles of terror.

They were dolphins. They circled around Kahu and seemed to be talking to her. She nodded and grabbed one around its body. As quick as a flash, the dolphins sped her to another area of the reef and stopped. Kahu seemed to say, 'Down here?' and the dolphins made a nodding motion.

Suddenly Kahu made a quick, darting gesture. She picked something up, inspected it, appeared satisfied with it, and went back to the dolphins. Slowly the girl and the dolphins rose towards us. But just as they were midway, Kahu stopped again. She kissed the dolphins goodbye and gave Nanny Flowers a heart attack by returning to the reef. She picked up a crayfish and resumed her upward journey. The dolphins were like silver dreams as they disappeared.

Nanny Flowers and I were treading water when Kahu appeared between us, smoothing her hair back from her face and blinking away the sea water. Nanny Flowers, sobbing, hugged her close in the water.

'I'm all right, Nanny,' Kahu laughed.

She showed the crayfish to us. 'This is for Paka's tea,' she said. 'And you can give him back his stone.'

She placed the stone in Nanny Flowers' hands. Nanny Flowers looked at me quickly. As we were pulling ourselves back into the dinghy she said, 'Not a word about this to Koro Apirana.'

I nodded. I looked back landward and in the distance saw the carving of Paikea on his whale like a portent.

As we got to the beach, Nanny Flowers said again, 'Not a word, Rawiri. Not a word about the stone or our Kahu.' She looked up at Paikea.

'He's not ready yet,' she said.

The sea seemed to be trembling with anticipation. Haumi e, hui e, *taiki e*.

Whale Song, Whale Rider

Fourteen

The muted thunder boomed under water like a great door opening far away. Suddenly the sea was filled with awesome singing, a song with eternity in it. Then the whale burst through the sea and astride the head was a man. He was wondrous to look upon. He was the whale rider.

He had come, the whale rider, from the sacred island far to the east. He had called to the whale, saying, 'Friend, you and I must take the gifts of life to the new land, mauri to make it fruitful.' The journey had been long and arduous, but the whale had been filled with joy at the close companionship they shared as they sped through the southern seas.

Then they had arrived at the land, and at a place called Whangara the golden rider had dismounted. He had taken the gifts of Hawaiki to the people and the land and sea had blossomed.

For a time the whale had rested in the sea which sighed at Whangara. Time had passed like a swift current, but in its passing had come the first tastes of separation. His golden master had met a woman and had married her. Time passed, time passed like a dream. One day, the whale's golden master had come to the great beast and there had been sadness in his eyes.

'One last ride, e hoa,' his master had said.

In elation, anger and despair, the whale had taken his golden master deeper than ever before and had sung to him of the sacred islands and of their friendship. But his master had been firm. At the end of the ride, he had said, 'I have been fruitful and soon children will come to me. My destiny lies here. As for you, return to the Kingdom of Tangaroa and to your own kind.'

The heartache of that separation had never left the whale, nor had the remembrance of that touch of brow to brow in the last hongi.

Antarctica. The Well of the World. Te Wai Ora o te Ao. Above, the frozen continent was swept with an inhuman, raging storm. Below, where the Furies could not reach, the sea was calm and unworldly. The light played gently on the frozen ice layer and bathed the undersea kingdom with an unearthly radiance. The giant roots of the ice extending down from the surface sparkled, glowed, twinkled and flashed prisms of light like strobes in a vast subterranean cathedral. The ice cracked, moaned, shivered and susurrated with rippling glissandi, a giant organ playing a titanic symphony.

Within the fluted ice chambers the herd of whales moved with infinite grace in holy procession. As they did so they offered their own choral harmony to the natural orchestration. Their movements were languid and lyrical, and belied the physical reality of their sizes; their tail flukes gently stroked the water, manoeuvring them ever southward. Around and above them the sealions, penguins and other Antarctic denizens darted, circled and swooped in graceful waltz.

Then the whales could go no further. Their sonics indicated that there was nothing in front except a solid wall of ice. Bewildered, the ancient bull whale let loose a ripple of harmonics, a plaintive cry for advice. Had his golden master been with him, he would have been given the direction in which to turn.

All of a sudden a shaft of light penetrated the underwater world and turned it into a gigantic hall of mirrors. In each one the ancient whale seemed to see a vision of himself being spurred ahead by his golden master. He made a quick turn and suddenly shards of ice began to cascade like spears around the herd. The elderly females throbbed their alarm to him. They were already further south than they had ever been before and the mirrors, for them, appeared only to reflect a crystal tomb for the herd. They communicated the urgency of the situation to their leader.

The aurora australis played above the ice world and the reflected light was like a mesmerising dream to the ancient bull whale. He began to follow the light, turning away from the southward plunge. As he did so he increased his speed, and the turbulence of his wake caused ice waterfalls within the undersea kingdom. Twenty metres long, he no longer possessed the flexibility to manoeuvre at speed.

The herd followed through the crashing, falling ice. They saw their leader rising to the surface and watched as the surface starred around him. They began to mourn, for they knew that *their journey to the dangerous islands was now a reality. Their leader was totally ensnared in the rhapsody of his dreams of the golden rider. So long part of their own whakapapa and legend, the golden rider could not be dislodged from their leader's thoughts. The last journey had begun and at the end of it Death was waiting.*

The aurora australis was like Hine Nui Te Po flashing above the radiant land. The whales swept swiftly through the southern seas.

Haumi e, hui e, taiki e.

Fifteen

Not long after Kahu's dive for the stone, in the early hours of the morning, a young man was jogging along Wainui Beach, not far from Whangara, when he noticed a great disturbance on the sea. 'The horizon all of a sudden got lumpy,' he said as he tried to describe the phenomenon, 'and lumps were moving in a solid mass to the beach.' As he watched, the jogger realised that he was witnessing the advance to the beach of a great herd of whales. 'They kept coming and coming,' he told the *Gisborne Herald*, 'and they didn't turn away. I ran down to the breakwater. All around me the whales were stranding themselves. They were whistling, an eerie, haunting sound. Every now and then they would spout. I felt like crying.'

The news was quickly communicated to the town, and the local radio and television stations sent reporters out to Wainui. One enterprising cameraman hired a local helicopter to fly him over the scene. It is his flickering film images that most of us remember. In the early morning light, along three kilometres of coastland, are two hundred whales, male, female, young, waiting to die. The waves break over them and hiss around their passive frames. Dotted on the beach are human shapes, drawn to the tragedy. The pilot of the helicopter says on camera, 'I've been to Vietnam, y'know, and I've done deer culling down south.' His lips are trembling and his eyes are moist with tears. 'But my oath, this is like seeing the end of the world.'

One particular sequence of the news film will remain indelibly imprinted on our minds. The camera zooms in on one of the whales, lifted high onto the beach by the waves. A truck has been driven down beside the whale. The whale is on its side, and blood is streaming from its mouth. The whale is still alive.

Five men are working on the whale. They are splattered with blood. As the helicopter hovers above them, one of the men stops his work and smiles directly into the camera. The look is triumphant. He lifts his arms in a victory sign and the camera sees that he has a chainsaw in his hand. Then the camera focusses on the other men, where they stand in the surging water. The chainsaw has just completed cutting through the whale's lower jaw. The men are laughing as they wrench the jaw from the butchered whale. There is a huge spout of blood as the jaw suddenly snaps free. The blood drenches the men in a dark gouting stream. Blood, laughing, pain, victory, blood.

It was that sequence of human butchery, more than any other, which triggered feelings of sorrow and anger among the people on the Coast. Some would have argued that in Maori terms a stranded whale was traditionally a gift from the Gods and that the actions could therefore be condoned. But others felt more primal feelings of aroha for the beasts which had once been our companions from the Kingdom of the Lord

Tangaroa. Nor was this just a question of one whale among many; this was a matter of two hundred members of a vanishing species.

At the time Kahu had just turned eight and Koro Apirana was down in the South Island with Porourangi. I rang them up to tell them what was happening. Koro said, 'Yes, we know. Porourangi has rung the airport to see if we can get on the plane. But the weather's cracked down on us and we can't get out. You'll have to go to Wainui. This is a sign to us. I don't like it. I don't like it at all.'

Luckily, knowing Kahu's kinship with the sea, I was glad that she had still been asleep when the news was broadcast. I said to Nanny Flowers, 'You better keep our Kahu at home today. Don't let her know what has happened.' Nanny's eyes glistened. She nodded her head.

I got on my motorbike and went round rousing the boys. I hadn't realised it before, but when you catch people unawares you sure find out a lot about them. For instance, one of the boys slept on his puku with his thumb in his mouth. Billy had his hair in curlers and he still had a smoke dangling from his lips. And a third slept with all his clothes on and the motorbike was in the bed with him.

'Come on, boys,' I said. 'We've got work to do.' We assembled at the crossroads, gunned our bikes, and then took off. Instead of going the long way by road we cut across country and beach, flying like mauri to help save the whales. The wind whistled among us as we sped over the landscape. Billy led the way and we followed — he was sure tricky, all right, knowing the shortcuts. No wonder the cops could never catch him. We flew over fences, jounced around paddocks, leapt streams and skirted the incoming tide. We were all whooping and hollering with the excitement of the ride when Billy took us up to a high point overlooking Wainui.

'There they are,' he said.

Gulls were wheeling above the beach. For as far as the eye could see whales were threshing in the curve of sand. The breakers were already red with blood. We sped down on our rescue mission.

The gulls cried, outraged, as we varoomed through their gathering numbers. The first sight to greet our eyes was this old pakeha lady who had sat down on a whale that some men were pulling onto the beach with a tractor. They had put a rope round the whale's rear flukes and were getting angrier and angrier with the woman, manhandling her away. But she would just return and sit on the whale again, her eyes glistening. We came to the rescue and that was the first punch-up of the day.

'Thank you, gentlemen,' the lady said. 'The whale is already dead of course, but how can men be so venal?'

By that time many of the locals were out on the beach. Some of them still had their pyjamas on. There were a lot of elderly people living near Wainui and it was amazing to see them trying to stop younger men from pillaging the whales. When one of the old women saw us, she set her mouth grimly and raised a pink slipper in a threatening way.

'Hey lady,' Billy said. 'We're the goodies.'

She gaped disbelievingly. Then she said, 'Well, if you're the goodies, you'd better go after them baddies.' She pointed down to where a truck was parked beside a dying whale. There were several beefy guys loading a dismembered jaw onto the back. As we approached we saw an old man scuffling with them. One of the young men smacked him in the mouth and the old man went down. His wife gave a high pipping scream.

We roared up to the truck.

'Hey, *man*,' I hissed. 'That whale belongs to Tangaroa.' I pointed to the dying beast. The stench of guts and blood was nauseous. Seagulls dived into the bloodied surf.

'Who's stopping us?'

'We are,' Billy said. He grabbed the chainsaw, started it up and, next minute, had sawn the front tyres of the truck. That started the second punch-up of the day.

It was at this stage that the police and rangers arrived. I guess they must have had trouble figuring out who were the goodies and who were the baddies because they started to manhandle us as well. Then the old lady with the pink slipper arrived. She waved it in front of the ranger and said, 'Not *them*, you stupid fool. They're on *our* side.'

The ranger laughed. He looked us over quickly. 'In that case, lady, I guess we'll have to work together. Okay, fellas?'

I looked at the boys. We had a strange relationship with the cops. But this time we nodded agreement.

'Okay,' the ranger said. 'The name's Derek. Let's get this beach cleared and cordoned off. We've got some Navy men coming in soon from Auckland.' He yelled, 'Anybody here got wet suits? If so, get into them. We'll need all the help we can get.'

The boys and I cleared the beach. We mounted a bike patrol

back and forth along the sand, keeping the spectators back from the water. The locals helped us. I saw a shape I thought I knew tottering down to the sea. The woman must have borrowed her son's wet suit, but I would have recognised those pink slippers anywhere.

All of us who were there that day and night will be forever bonded by our experience with the stranded whales. They were tightly bunched and they were crying like babies. Derek had assigned people in groups, eight people to look after each whale. 'Try to keep them cool,' he said. 'Pour water over them, otherwise they'll dehydrate. The sun's going to get stronger. Keep pouring that water, but try to keep their blowholes clear — otherwise they'll suffocate to death. Above all, try to stop them from lying on their sides.'

It was difficult and heavy work, and I marvelled at the strength that some of the elderly folk brought to the task. One of the old men was talking to his whale and said in response to his neighbour, 'Well, *you* talk to your plants!' At that point the whale lifted its head and staring at the two men, gave what appeared to be a giggle. 'Why, the whale understands,' the old man said. So the word went down the line of helpers.

Talk to the whales.
They understand.
They understand.

The tide was still coming in. The Navy personnel arrived and members of Greenpeace, Project Jonah and Friends of the

Earth also. Two helicopters whirred overhead, dropping wet-suited men into the sea.

A quick conference was called on the situation. The decision was made to try and tow the whales out to sea. Small runabouts were used, and while most of the whales resisted being towed by the tail, there were some successes. In that first attempt, a hundred and forty whales were refloated. There were many cheers along the beach. But the whales were like confused children, milling and jostling out in the deeper water, and they kept trying to return to those who were still stranded along the beach, darting back to those who were already dead. The cheers became ragged when all the whales returned to beach themselves again at low tide.

'Okay, folks,' Derek called. 'We're back at the beginning. Let's keep them cool. And let's keep our spirits up.'

The sea thundered through his words. The seagulls screamed overhead. The sun reached noon and began its low decline. I saw children coming from buses to help. Some schools had allowed senior students to aid the rescue. Many of the old folk were pleased to be relieved. Others, however, stayed on. For them, *their* whale had become a member of the family. 'And I can't leave Sophie now,' an elderly lady said. The sun scattered its spokes across the sand.

The whales kept dying. As each death occurred the people who were looking after the whale would weep and clasp one another. They would try to force away the younger, healthier whales which had returned to keep company with their dying mates. When a large whale was turning on its side, several juveniles would try to assist it, rubbing their bodies against the dying whale's head. All the time the animals were uttering cries of distress or alarm, like lost children.

Some old people refused to leave the beach. They began to sing 'Onward Christian Soldiers'. They continued to try to right the whales, rocking them back and forth to restore their balance, and encouraging them to swim in groups. It was soon obvious, however, that the whales did not wish to be separated. So the ranger decided that an effort should be made to herd the surviving whales as one large group out to sea. They seemed to sense that we were trying to help them and offered no resistance or harm. When we reached them, most were exhausted, but when they felt us lifting them up and pushing them out to sea they put their energy into swimming and blowing.

Somehow we managed to get the whales out again with the incoming tide. But all they did was to cry and grieve for their dead companions; after wallowing aimlessly, they would return to nuzzle their loved ones. The sea hissed and fell, surged and soughed upon the sand. The whales were singing a plaintive song, a fluting sound which began to recede away, away, away.

By evening, all the whales had died. Two hundred whales, lifeless on the beach and in the water. The boys and I waited during the death throes. Some of the people from the town had set up refreshments and were serving coffee. I saw the lady with the pink slippers sipping coffee and looking out to sea.

'Remember me?' I said. 'My name's Rawiri. I'm a goodie.'

There were tears in her eyes. She pressed my hand in companionship.

'Even the goodies,' she said, 'can't win all the time.'

When I returned to Whangara that night, Nanny Flowers said, 'Kahu knows about the whales.' I found Kahu way up on the bluff, calling out to sea. She was making that mewling sound and then cocking her head to listen for a reply. The sea was silent, eternal.

I comforted her. The moon was drenching the sky with loneliness. I heard an echo of Koro Apirana's voice, 'This is a sign to us. I don't like it.' Suddenly, with great clarity, I knew that our final challenge was almost upon us. I pressed Kahu close to me, to reassure her. I felt a sudden shiver as far out to sea, muted thunder boomed like a door opening far away.

Haumi e, hui e, *taiki e*.

Sixteen

Yes, people in the district vividly remember the stranding of the whales because television and radio brought the event into our homes that evening. But there were no television cameras or radio newsmen to see what occurred in Whangara the following night. Perhaps it was just as well, because even now it all seems like a dream. Perhaps, also, the drama enacted that night was only meant to be seen by the iwi and nobody else. Whatever the case, the earlier stranding of whales was merely a prelude to the awesome event that followed, an event that had all the cataclysmic power and grandeur of a Second Coming.

The muted thunder and forked lightning the day before had advanced quickly across the sea like an illuminated cloud. We saw it as a great broiling rush of the elements; with it came the icy cold winds hurled from the Antarctic.

Nanny Flowers, Kahu and I were watching the weather anxiously. We were at the airport, waiting for the flight bringing Koro Apirana and Porourangi back to us. Suddenly there was the plane, bucking like an albatross, winging ahead of winds which heralded the arrival of the storm. It was as if

107

Tawhirimatea was trying to smash the plane down to earth in anger.

Koro Apirana was pale and upset. He and Nanny Flowers were always arguing, but this time he was genuinely relieved to see her. 'Taku Putiputi,' he whispered as he held her tightly.

'We had a hard time down South,' Porourangi said, trying to explain Koro's agitation. 'The land dispute was a difficult one and I think that Koro is worried about the Judge's decision. Then when he heard about the whales, kei te pouri ia, he grew very sombre.'

The wind began to whistle and shriek like wraiths.

'Something's going on,' Koro Apirana whispered. 'I don't know what it is. But something —'

'It's all right,' Kahu soothed. 'It will be all right, Paka.'

We collected the suitcases and ran out to the station wagon. As we drove through the town the illuminated cloud seemed always to be in front of us, like a portent.

Even before we reached Wainui Beach we could smell and taste the Goddess of Death. The wind was still lashing like a whip at the landscape. The car was buffeted strongly, and Nanny Flowers was holding on to her seat belt nervously.

'It's all right,' Kahu said. 'There, there, Nanny.'

Suddenly, in front of the car, I could see a traffic officer waving his torch. He told us to drive carefully as earth-moving machinery was digging huge trenches in the sand for the dead whales. Then he recognised me as one of the people who had tried to help. His smile and salute were sad.

I drove carefully along the highway. On our right I could see the hulking shapes of the graders, silhouetted against the broiling sky. Further down the beach, at the ocean's edge, were the whales, rocked by the surge and hiss of the sea. The whole scene was like a surreal painting, not nightmarish, but immensely tragic. What had possessed the herd to be so suicidal? The wind hurled sand and mud at the windscreen of the station wagon. We watched in silence.

Then, 'E tu,' Koro Apirana said.

I stopped the station wagon. Koro Apirana got out. He staggered against the onslaught of the wind.

'Leave him,' Nanny Flowers said. 'Let him be with the whales by himself. He needs to mourn.'

But I was fearful of Koro's distraught state. I got out of the car too. The wind was freezing. I walked over to him. His eyes were haunted. He looked at me, uncomprehending.

'No wai te he?' he shouted. 'Where lies the blame?'

And the seagulls caught his words within their claws and screamed and echoed the syllables overhead.

When we turned back to the station wagon I saw Kahu's white face, so still against the window.

'This is a sign to us,' Koro Apirana said again.

We turned off the Main Highway and onto the road to Whangara. It was so dark that I switched the headlights on full. I looked up at that illuminated cloud. I had the strangest feeling that its centre was just above the village. I felt a rush of fear and was very glad when Whangara came into sight.

Whangara must be one of the most beautiful places in the world, like a kingfisher's nest floating on the water at summer solstice. There it was, with church in the foreground and marae behind, silhouetted against the turbulent sea. And there was Paikea, our eternal sentinel, always vigilant against any who would wish to harm his descendants. Caught like this, the village was a picture of normality given the events that were to come.

I drove up to the homestead.

'Kahu, you help take Koro's bags inside,' I said. 'Then I'll take you and Daddy home. Okay?'

Kahu nodded. She put her arms around her grandfather and said again, 'It's all right, Paka. Everything will be all right.'

She picked up a small flight bag and carried it up onto the verandah. We were all getting out of the wagon and climbing towards her when, suddenly, the wind died away.

I will never forget the look on Kahu's face. She was gazing out to sea and it was as if she was looking back into the past. It was a look of calm, of acceptance. It forced us all to turn to see what Kahu was seeing.

The land sloped away to the sea. The surface of the water was brilliant green, blending into dark blue and then a rich purple. The illuminated cloud was seething above one place on the horizon.

All of a sudden there was a dull booming from beneath the water, like a giant door opening a thousand years ago. At the place below the clouds the surface of the sea shimmered like gold dust. Then streaks of blue lightning came shooting out of the sea like missiles. I thought I saw something flying through the air, across the aeons, to plunge into the marae.

A dark shadow began to ascend from the deep. Then there were other shadows rising, ever rising. Suddenly the first shadow breached the surface and I saw it was a whale. Leviathan. Climbing through the pounamu depths. Crashing through the skin of sea. And as it came, the air was filled with streaked lightning and awesome singing.

Koro Apirana gave a tragic cry, for this was no ordinary beast. This was a tipua. A taniwha. As it came it filled the air with its singing.

Karanga mai, karanga mai,
karanga mai.

Its companions began to breach the surface also, orchestrating the call with unearthly music.

The storm finally unleashed its fury and strength upon the land. The sea was filled with whales and in their vanguard was their ancient battle-scarred leader.

Karanga mai, karanga mai,
karanga mai.

On the head of the whale was the sacred sign. A swirling moko, flashing its mana across the darkening sky.

I zoomed on my bike through the night and the rain, rounding the boys up. 'I'm sorry, boys,' I said to them as I yanked them out of bed, 'we're needed again.'

'Not more whales,' they groaned.

'Yes,' I said. 'But this is different boys, different. These whales are right here in Whangara.'

Koro Apirana had issued his instructions to Porourangi and me. We were to gather up the boys and all the available men of the village, and tell them to come to the meeting house. And we were to hurry.

'Huh?' Nanny Flowers had said in a huff. 'What about us women! We've got hands to help.'

Koro Apirana smiled a wan smile. His voice was firm as he told her, 'I don't want you to interfere, e Kui. You know as well as I do that this is tapu work.'

Nanny Flowers bristled. 'But you haven't got enough men to help. You watch out. If I think you need the help, well, kia whakatane ake au i ahau. Just like Muriwai, koroua.'

'In the meantime,' Koro Apirana said, 'you leave the organising to me. If the women want to help, you tell them to meet you in the dining room. I'll leave them to you.'

He kissed her and she looked him straight in the eyes. 'I say again,' she warned, 'I'll be like Muriwai if I have to. Kahu, also, if she has to be.'

'You keep Kahu away, e Kui,' Koro Apirana said. 'Porearea ia. She's of no use to me.'

With that he had turned to Porourangi and me. As for Kahu, she was staring at the floor, resigned, feeling sorry for herself.

Together, we had all watched the whale with the sacred sign plunging through the sea towards us. The attending herd had fallen back, sending long undulating calls to the unheeding bull whale, which had propelled itself forcefully onto the beach. We had felt the tremor of its landing. As we watched, fearfully, we saw the bull whale heaving itself by muscle contraction even further up the sand. Then, sighing, it had rolled onto its right side and prepared itself for death.

Five or six elderly females had separated from the herd to lie close to the bull whale. They sang to it, attempting to encourage it back to the open sea where the rest of the herd were waiting. But the bull whale remained unmoving.

We had run down to the beach. None of us had been prepared for the physical size of the beast. It seemed to tower over us. A primal psychic force gleamed in its swirling moko. Twenty metres long, it brought with it a reminder of our fantastic past.

Then, in the wind and the rain, Koro Apirana had approached the whale. 'E te tipua,' he had called, 'tena koe. Kua tae mai koe ki te mate? Ara, ki te ora.' There had been no reply to his question: Have you come to die, or to live? But we had the feeling that this was a decision which had been placed in *our* hands. The whale had raised its giant tail fin:

That is for you to decide.

It was then that Koro Apirana had asked that the men gather in the meeting house.

Outside there was wind and rain, lightning and thunder. The lightning lit up the beach where the stranded whale was lying. Far out to sea the whale herd waited, confused. Every now and then one of the elderly females would come to hongi the ancient whale and to croon its aroha for him.

Inside the stomach of the meeting house there was warmth, bewilderment, strength and anticipation, waiting to be soldered into a unity by the words of our rangatira, Koro Apirana. The sound of the women assembling in the wharekai under Nanny Flowers' supervision came to us like a waiata aroha. As I shut the door to the meeting house I saw Kahu's face, like a small dolphin, staring out to sea. She was making her mewling noise.

Koro Apirana took us for karakia. His voice rose and fell like the sea. Then he made his greetings to the house, our ancestors, and the iwi gathered inside the house. For a moment he paused, searching for words to begin his korero.

'Well, boys,' he said, 'there are not many of us. I count twenty-six —'

'Don't forget me, Koro,' a six-year-old interjected.

'Twenty-seven, then,' Koro Apirana smiled, 'so we all have to be one in body, mind, soul and spirit. But first we have to agree on what we must do.' His voice fell silent. 'To explain, I have to talk philosophy and I never went to no university. My university was the school of hard knocks —'

'Kia kaha, koro. Kia kaha,' someone yelled.

'So I have to explain in my own way. Once, our world was one where the Gods talked to our ancestors and man talked with the Gods. Sometimes the Gods gave our ancestors special powers. For instance, our ancestor Paikea' — Koro Apirana gestured to the apex of the house — 'was given power to talk to whales and to command them. In this way, man, tipua and Gods lived in close communion with one another.'

Koro Apirana took a few thoughtful steps back and forward.

114

'But then,' he continued, 'man assumed a cloak of arrogance and set himself up above the Gods. He even tried to defeat Death, but failed. As he grew in his arrogance he started to drive a wedge through the original oneness of the world. In the passing of Time he divided the world into that half he could believe in and that half he could not believe in. The real and the unreal. The natural and supernatural. The present and the past. The scientific and the fantastic. He put a barrier between both worlds and everything on his side was called rational and everything on the other side was called irrational. Belief in our *Maori* Gods,' he emphasised, 'has often been considered irrational.'

Koro Apirana paused again. He had us in the palms of his hands and was considerate about our ignorance, but I was wondering what he was driving at. Suddenly he gestured to the sea.

'You have all seen the whale,' he said. 'You have all seen the sacred sign tattooed on its head. Is the moko there by accident or by design? Why did a whale of its appearance strand itself here and not at Wainui? Does it belong in the real world or the unreal world?'

'The real,' someone called.

'Is it natural or supernatural?'

'It is supernatural,' a second voice said.

Koro Apirana put up his hands to stop the debate. 'No,' he said, 'it is *both*. It is a reminder of the oneness which the world once had. It is the pito joining past and present, reality and fantasy. It is both. It is *both*,' he thundered, 'and if we have forgotten the communion then we have ceased to be Maori.'

The wind whistled through his words. 'The whale is a sign,' he began again. 'It has stranded itself here. If we are able to return it to the sea, then that will be proof that the oneness is still with us. If we are not able to return it, then this is because we have become weak. If it lives, we live. If it dies, we die. Not only its salvation but ours is waiting out there.'

Koro Apirana closed his eyes. His voice drifted in the air and hovered, waiting for a decision.

'Kei te ora tatou? Ara, kei te mate.'

Our answer was an acclamation of pride in our iwi.

Koro Apirana opened his eyes. 'Okay then, boys. Let's go down there and get on with it.'

Porourangi gave the orders. He told the men that they were to drive every available truck, car, motorbike or tractor down to the bluff overlooking the sea and flood the beach with their headlights. Some of the boys had spotlights which they used when hunting opossums; these, also, were brought to the bluff and trained on the stranded whale. In the light, the moko flared like a silver scroll.

Watching from the wharekai, Nanny Flowers saw Koro Apirana walking around in the rain and got her wild up. She yelled out to one of the boys, '*Hoi*, you take his raincoat to that old paka. Thinks he's Super Maori, ne.'

'What are they doing, Nanny?' Kahu asked.

'They're taking all the lights down to the beach, mokopuna,' Nanny Flowers answered. 'Our tipua must be returned to the sea.'

Kahu saw the beams from the headlights of two tractors

cutting through the dark. Then she saw her father, Porourangi, and some of the boys running down to the whale with ropes in their hands.

'That's it, boys,' Koro Apirana yelled. 'Now who are the brave ones to go out in the water and tie the ropes around the tail of our tipua? We have to pull him around so that he's facing the sea. Well?'

I saw my mate, Billy, and volunteered on his behalf.

'Gee thanks, pal,' Billy said.

'I'll take the other rope,' Porourangi offered.

'No,' Koro Apirana said. 'I need you here. Give the rope to your brother, Rawiri.'

Porourangi laughed and threw the rope to me. 'Hey, I'm not your brother,' I said.

He pushed me and Billy out into the sea. The waves were bitingly cold and I was greatly afraid because the whale was so gigantic. As Billy and I struggled to get to the tail all I could think of was that if it rolled I would be squashed just like a nana. The waves lifted us up and down, up and down, up into the dazzle of the lights on the beach and down into the dark sea. Billy must have been as frightened of the whale as I was because he would say, 'Excuse me, koro,' whenever a wave smashed him into the side of the whale, or, 'Oops, sorry koro.'

'Kia tere! Kia tere!' Koro Apirana yelled from the beach, 'We haven't much time. Stop mucking around.'

Billy and I finally managed to get to the tail of the tipua. The flukes of the whale were enormous, like huge wings.

'One of us will have to dive underneath,' I suggested to Billy, 'to get the ropes around.'

'Be my guest,' Billy said. He was hanging on for dear life.

There was nothing for it but to do the job myself. I took three deep breaths and dived. The water was churning with sand and small pebbles and I panicked when the whale moved. Just my luck if it did a tiko, I thought. I sought the surface quickly.

'E ora ana,' Billy shouted in triumph. I passed the two ropes to him. He knotted them firmly and we fought our way back to the beach. The boys gave a big cheer. I heard Billy boasting about how *he* had done all the hard work.

'Now what?' Porourangi asked Koro Apirana.

'We wait,' said Koro Apirana, 'for the incoming tide. The tide will help to float our tipua and, when he does we'll use the tractors to pull him around. We will only have the one chance. Then once he's facing the sea we'll all have to get in the water and try to push him out.'

'We could pull him out by boat,' I suggested.

'No, too dangerous,' Koro replied. 'The sea is running too high. The other whales are in the way. No, we wait. And we pray.'

Koro Apirana told Billy and I to get out of our wet clothes. We hopped on my motorbike and went up to the homestead to change. Naturally, Nanny Flowers with her hawk eyes saw us and came ambling over to ask what was happening down on the beach.

'We're waiting for the tide,' I said.

I thought that Nanny Flowers would start to growl and protest about not being involved. Instead she simply hugged me and said, 'Tell the old paka to keep warm. I want him to come back to me in one piece.'

Then Kahu was there, flinging herself into my arms. 'Paka? Is Paka all right?'

'Ae, Kahu,' I said.

Suddenly the horns of the cars down on the beach began to sound. The tide was turning. Billy and I rushed to the motor-bike and roared back.

'There, there,' Kahu said to Nanny Flowers. 'They'll be all right.'

By the time we got back to Koro Apirana the boys were already in action. 'The sea came up so sudden,' Porourangi yelled above the waves. 'Look.'

The whale was already half submerged, spouting in its distress. Three elderly females had managed to come beside him and were trying to nudge him upright before he drowned.

'*Now*,' Porourangi cried. The two tractors coughed into life. The rope took up between them and the whale, and quickly became taut.

With a sudden heave and suck of sand the whale gained its equilibrium. Its eyes opened, and Koro Apirana saw the mana and the wisdom of the ages shining like a sacred flame. The moko of the whale too seemed alive with unholy fire.

Ka ora tatou?

'E te tipua,' Koro Apirana said. 'Ae, ka ora tatou. Haere atu koe ki te moana. Me huri koe ki te Ao o Tangaroa.'

The tractors began to pull the whale round. By degrees it

119

was lying parallel to the beach. The boys and I put our shoulders to its gigantic bulk and tried to ease it further seaward.

It was then that the ropes snapped. Koro Apirana gave a cry of anguish, burying his face in his hands. Swiftly he turned to me. 'Rawiri, go tell your Nanny Flowers it is time for the women to act the men.'

Even before I reached the dining hall Nanny Flowers was striding through the rain. The women were following behind her.

'In we go, girls,' Nanny Flowers said. 'Kahu, you stay on the beach.'

'But Nanny.'

'E noho,' Nanny Flowers ordered.

The women ran to join us. Porourangi began to chant encouragement. 'Toia mai,' he called. 'Te waka,' we responded. 'Ki te moana,' he cried. 'Te waka,' we answered again. 'Ki Tangaroa,' he chanted. 'Te waka,' we replied a third time. And at each response we put our shoulders to the whale, pushing it further seaward and pointing it at the ocean stars.

Out to sea the herd sang its encouragement. The elderly females spouted their joy.

Kua mate, pea? A ripple ran along the back of the whale. A spasm. Our hearts leapt with joy. Suddenly the huge flukes rose to stroke at the sky.

The whale moved.

But our joy soon turned to fear. Even as the whale moved, Koro Apirana knew we had lost. For instead of moving out to sea the whale turned on us. The tail crashed into the water causing us to move away, screaming our dread. With a terrifying guttural moan the whale sought deeper water where we could not reach it. *Kei te mate*. Then, relentlessly, it turned shoreward again, half-submerging itself in the water, willing its own death.

Kei te mate.

The wind was rising. The storm was raging. The sea stormed across the sky. We watched, forlorn, from the beach.

'Why?' Kahu asked Koro Apirana.

'Our tipua wants to die.'

'But why?'

'There is no place for it here in this world. The people who once commanded it are no longer here.' He paused. 'When it dies, we die. I die.'

'*No*, Paka. And if it lives?'

'Then we live also.'

Nanny Flowers cradled the old man. She started to lead him away and up to the homestead. The sky forked with lightning. The iwi watched in silence, waiting for the whale to die. The elderly females cushioned it gently in its last resting place. Far out to sea the rest of the herd began the mournful waiata for their leader.

Seventeen

Nobody saw her slip away and enter the water. Nobody knew at all until she was half way through the waves. Then the headlights and spotlights from the cars along the beach picked up her white dress and that little head bobbing up and down in the waves. As soon as I saw her, I knew it was Kahu.

'Hey!' I yelled. I pointed through the driving rain. Other spotlights began to catch her. In that white dress and white ribboned pigtails she was like a small puppy, trying to keep its head up. A wave would crash over her but somehow she would appear on the other side, gasping wide-eyed, and doing what looked like a cross between a dog paddle and a breast stroke.

Instantly I ran through the waves. People said I acted like a maniac. I plunged into the sea.

If the whale lives, we live. These were the only words Kahu could think of.

We have lost our way of talking to whales.

The water was freezing, but not to worry. The waves were huge, but kei te pai. The rain was like spears, but hei aha.

Every now and then she had to take a deep breath because sometimes the waves were like dumpers, slamming her down to the sandy bottom, but somehow she bobbed right back up like a cork. Now, the trouble was that the lights from the beach were dazzling her eyes, making it hard to see where she was going. Her neck was hurting with the constant looking up, but *there, there*, was the whale with the moko. She dog paddled purposefully toward it. A wave smashed into her and she swallowed more sea water. She began to cough and tread water. Then she set her face with determination. As she approached the whale, she suddenly remembered what she should do.

That damn kid, I swore as I leapt into the surf. For one thing I was no hero and for another I was frightened by the heavy seas. Bathtubs were really the closest I ever liked to get to water and at least in a bath the water was hot. This wasn't. It was cold enough to freeze a person. I knew, because I'd only just before been in it.

But I had to admire the kid. She'd always been pretty fearless. Now, here she was, swimming towards the whale. I wondered what on earth she expected to do.

I saw Porourangi running after Koro Apirana and Nanny Flowers to bring them back. Then the strangest thing happened. I heard Kahu's high treble voice shouting something to the sea. She was singing to the whale. Telling it to acknowledge her coming.

'Karanga mai, karanga mai, karanga mai.' She raised her head and began to call to the whale.

The wind snatched at her words and flung them with the foam to smash in the wind.

Kahu tried again. 'E te tipua rangatira,' she called. 'Kei te haere atu au ki a koe. Ko Kahu ahau. Ko Kahutia Te Rangi ahau.'

The headlights and spotlights were dazzling upon the whale. It may have been the sudden light, or a cross current, but the eye of the whale seemed to flicker. Then the whale appeared to be looking at the young girl swimming.

Ko Kahutia Te Rangi?

'Kahu!' I could hear Nanny Flowers screaming in the wind.

My boots were dragging me down. I had to stop and reach under to take them off. I lost valuable time, but better that than drown. The boots fell away into the broiling currents.

I looked up. I tried to see where Kahu was. The waves lifted me up and down.

'Kahu, no,' I cried.

She had reached the whale and was hanging onto its jaw. 'Tena koe e koro,' Kahu said as she clung onto the whale's jaw. 'Tena koe.' She patted the whale and looking into its eye, said, 'Kei te haramai ahau ki a koe.'

The swell lifted her up and propelled her away from the head of the whale. She choked with the water and tried to dog paddle back to the whale's eye.

'Help me,' she cried. 'Ko Kahutia Te Rangi au. Ko Paikea.'

The whale shuddered at the words.

By chance, Kahu felt the whale's forward fin. Her fingers tightened quickly around it. She held on for dear life.

And the whale felt a surge of gladness which, as it mounted, became ripples of ecstasy, ever increasing. He began to communicate his joy to all parts of his body.

Out beyond the breakwater the herd suddenly became alert. With hope rising, they began to sing their encouragement to their leader.

'Kahu no,' I cried again. I panicked and I lost sight of her, and I thought that she had been swept into the whale's huge mouth. I was almost sick thinking about it, but then I remembered that Jonah had lived on in the belly of his whale. So, if necessary, I would just have to go down *this* whale's throat and pull Kahu right out. No whale was going to swallow our Kahu and get away with it.

The swell lifted me up again. With relief I saw that Kahu was kei te pai. She was hanging onto the whale's forward fin. For a moment I thought my imagination was playing tricks. Earlier, the whale had been lying on its left side. But now it was righting itself, rolling so that it was lying on its stomach.

Then I felt afraid that in the rolling Kahu would get squashed. No, she was still hanging onto the fin. I was really frightened though, because in the rolling Kahu had been lifted clear of the water and was now dangling on the side of the whale, like a small white ribbon.

The elderly female whales skirled their happiness through

the sea. They listened as the pulsing strength of their leader manifested itself in stronger and stronger whalesong. They crooned tenderness back to him and then throbbed a communication to the younger males to assist their leader. The males arranged themselves in arrow formation to spear through the raging surf.

'Kia ora, e koro,' Kahu whispered. She was cold and exhausted. She pressed her cheek to the whale's side and kissed the whale. The skin felt like very smooth, slippery rubber.

Without really thinking about it, Kahu began to stroke the whale just behind the fin. *Taku rangatira*. She felt a tremor in the whale and a rippling under the skin. Suddenly she saw that indentations like footholds and handholds were appearing before her. She tested the footholds and they were firm. Although the wind was blowing fiercely she stepped away from the sheltering fin and began to climb. As she did so, she caught a sudden glimpse of her Koro Apirana and Nanny Flowers clustered with the others on the faraway beach.

I was too late. *Taku hoa*. I saw Kahu climbing the side of the whale. A great wave bore me away from her. I yelled out to her, a despairing cry.

Kahu could climb no further. *Ko Kahutia Te Rangi*. She saw the rippling skin of the whale forming a saddle with fleshy stirrups for her feet and pommels to grasp. She wiped her eyes and smoothed down her hair as she settled herself astride the whale. She heard a cry, like a moan in the wind.

I saw black shapes barrelling through the breakers. Just my luck, I thought. If I don't drown I'll get eaten.

Then I saw that the shapes were smaller whales of the herd, coming to assist their leader.

The searchlights were playing on Kahu astride the whale. *Taku tangata*. She looked so small, so defenceless.

Quietly, Kahu began to weep. She wept because she was frightened. She wept because Paka would die if the whale died. She wept because she was lonely. She wept because she loved her baby sister and her father and Ana. She wept because Nanny Flowers wouldn't have anyone to help her in the kumara patch. She wept because Koro Apirana didn't love her. And she also wept because she didn't know what dying was like.

Then, screwing up her courage, she started to kick the whale as if it was a horse.

'Haere tohora, haere,' she shrilled.

The whale began to rise in the water.

'E huri, e huri ki te moana,' she cried.

Slowly, the whale began to turn to the open sea. *Ae, taku rangatira*. As it did so, the younger whales came to push their leader into deeper water.

'Ka haere atu maua,' she ordered.

Together, the ancient whale and its escort began to swim into the deep ocean.

She was going, our Kahu. She was going into the deep ocean. I could hear her small piping voice in the darkness as she left us.

She was going with the whales into the sea and the rain. She was a small figure in a white dress, kicking at the whale as if it was a horse, her braids swinging in the rain. Then she was gone and we were left behind.

Ko Paikea, ko Paikea.

Eighteen

She was the whale rider. Astride the whale she felt the sting of the surf and rain upon her face. On either side the younger whales were escorting their leader through the surf. They broke through into deeper water.

Her heart was pounding. She saw that now she was surrounded by the whale herd. Every now and then, one of the whales would come to rub alongside the ancient leader. Slowly, the herd made its way to the open sea.

She was Kahutia Te Rangi. She felt a shiver running down the whale and, instinctively, she placed her head against its skin and closed her eyes. The whale descended in a shallow dive and the water was like streaming silk. A few seconds later the whale surfaced, gently spouting.

Her face was wet with sea and tears. The whales were gathering speed, leaving the land behind. She took a quick look and saw headlights far away. Then she felt that same shiver again, and again, placed her head against the whale's skin. This time when the whale dived, it stayed underwater longer. But Kahu had made a discovery. Where her face was pressed the whale had opened up a small breathing chamber.

She was Paikea. In the deepening ocean the fury of the storm was abating. The whale's motions were stronger. As it rose from the sea, its spout was a silver jet in the night sky. Then it dived a third time, and the pressure on her eardrums indicated to the young girl that this was a longer dive than the first two had been. And she knew that the next time would be forever.

She was serene. When the whale broke the surface she made her karanga to sky and earth and sea and land. She called her farewells to her iwi. She prepared herself as best she could with the little understanding she had. She said goodbye to her Paka, her Nanny, her father and mother, her Uncle Rawiri, and prayed for their good health always. She wanted them to live for ever and ever.

The whale's body tensed. The girl felt her feet being locked by strong muscles. The cavity for her face widened. The wind whipped at her hair.

Suddenly the moon came out. Around her the girl could see whales sounding, sounding, sounding. She lowered her face into the whale and closed her eyes. 'I am not afraid to die,' she whispered to herself.

The whale's body arched and then slid into a steep dive. The water hissed and surged over the girl. The huge flukes seemed to stand on the surface of the sea, stroking at the rain-drenched sky. Then slowly, they too slid beneath the surface.

She was Kahutia Te Rangi. She was Paikea. She was the whale rider.

Hui e, haumi e, *taiki e*.

The iwi were weeping on the beach. The storm was leaving with Kahu. Nanny Flowers' heart was racing and her tears were streaming down her face. She reached into her pockets for a handkerchief. Her fingers curled around a carved stone. She took it out and gave it to Koro Apirana.

'Which of the boys?' he gasped in grief. 'Which of the —'

Nanny Flowers was pointing out to sea. Her face was filled with emotion as she cried out to Kahu. The old man understood. He raised his arms as if to claw down the sky upon him.

The Girl from the Sea

Nineteen

Apotheosis. In the *sunless sea sixty whales were sounding slowly, steeply diving. An ancient bull whale, twenty metres long and bearing a sacred sign, was in the middle of the herd. Flanking him were seven females, half his size, like black-gowned kuia, shepherding him gently downward.*

'Haramai, haramai e koro,' *the kuia sibilantly sang.* 'Tomo mai i waenganui i o tatou iwi.'

The sea hissed and sparkled with love for the old koroua and, every now and then, the kai karanga among the kuia would close in on him, gently, to nuzzle him, caress him, and hongi him just to let him know how much he had been missed. But in her heart of hearts she knew that he was badly wounded and near to exhaustion.

From the corner of her eye, the kai karanga noticed a small white shape clasping the koroua just behind his moko. She rose to observe the figure and then drifted back beside him.

'Ko wai te tekoteko kei runga?' *she sang, her voice musically pulsing.*

'Ko Paikea, ko Paikea,' *the koroua responded, and the bass notes boomed like an organ through the subterranean cathedral of the sea.*

*T*he sea was a giant liquid sky and the whales were descending, plummeting downward like ancient dreams. On either side of the koroua and his kuia entourage were warrior whales, te hokowhitu a Tu, swift and sturdy, always alert, a phalanx of fierceness.

'Neke neke,' the toa taiaha warned. 'Neke neke.'

He signalled to some of the warriors to fall back to the rear to close up and tighten the remaining ope, kui ma, koro ma, tamariki ma.

The kai karanga was processing the information that the koroua had given her. 'Ko Paikea? Ko Paikea?' The other kuia caught flashes of her puzzlement and, curious themselves, rose to look at the motionless rider. One of them nudged the tiny shape and saw a white face like a sleeping dolphin. The kuia hummed their considerations among themselves, trying to figure it all out. Then they shrugged. 'Aua.' If the koroua said it was Paikea, it was Paikea. After all, the koroua was the boss, the rangatira, and e hika they knew how crotchety he became if they did not respect his mana.

'Neke neke,' the toa taiaha whistled reprovingly.

The whales shifted closer together, to support one another, as they fell through the sea.

'*K*o Paikea? Ko Paikea?' the kai karanga wondered anxiously. Although the kai karanga loved the koroua, and had done so for many whaleyears, she was not blind to his faults. Over the last few years, for instance, the koroua had become more and more depressed, considering that death was upon him and revisiting the places of his memory. The Valdes Peninsula.

Tonga. Galapagos. Tokelau. Easter Island. Rarotonga. Hawaiki, the Island of the Ancients. Antarctica. Now, Whangara, where he had almost been lost to the iwi.

Then she knew.

'E tu,' the kai karanga called. In her memory's eye she saw Paikea himself and he was flinging small spears seaward and landward.

Instantly the ope ceased its sounding and became poised in mid flight between the glassy surface of the sea and the glittering ocean abyss.

The toa taiaha glided up to the kai karanga. 'He aha te mate?' he trumpeted belligerently. The kai karanga was always calling for a halt.

The kai karanga's heart was pounding. 'Kei te korero ahau,' she said sweetly, 'ki te koroua.' So saying, she descended gently toward the koroua, to talk with him.

*T*he sea scintillated with the sweetness of the kai karanga as she hovered near her ancient mate. Illuminated jellyfish exploded silvered starbursts through the dark depths. Far below, a river of phosphorescence lent lambent light to the abyss like a moonlit tide. The ocean was alive with noises: dolphin chatter, krill hiss, squid thresh, shark swirl, shrimp click and, ever present, the strong swelling chords of the sea's constant rise and fall.

'E koro,' the kai karanga began in a three-tone sequence drenched with love. 'E taku rangatira,' she continued, adding a string of harmonics. 'E taku tane,' she breathed with slyness, threading her words with sensuous major arpeggios, 'e hara tera tekoteko ko Paikea.'

The other kuia edged away carefully but they secretly admired the courage of the kai karanga in questioning the identity of the tekoteko kei runga.

'Ko Paikea,' the koroua said, insistent, 'ko Paikea.'

The kai karanga cast her eyes downward, hoping that the koroua would take this as a sign of feminine submission, but she knew in fact what she was up to.

'Kaore, kaore,' she belled sweetly.

The kuia gasped at the kai karanga's stubbornness. The toa taiaha waited for the word from his leader to teach her a lesson.

The koroua responded in a testy manner. 'Kei te korero tenei tekoteko ki ahau,' he said. 'Tana ingoa ko Kahutia Te Rangi.' He was recalling that when the tekoteko spoke to him, it had been to say that its name was Kahutia Te Rangi and, as the kai karanga should know, this was another name for Paikea. 'Ko Kahutia Te Rangi ko Paikea.'

The kai karanga allowed herself to drift just below the koroua.

'Pea, pea,' the kai karanga trilled in soprano tones of innocent guile.

The other kuia now decided to give the kai karanga a wide berth. She had a lot of gumption, all right. Fancy saying, 'Perhaps,' to the koroua.

The kai karanga saw the toa taiaha preparing to give her a sharp nip in the behind. She moved quickly toward the koroua and let a fin accidentally on purpose caress the place of his deepest pleasure. 'Engari,' she told the koroua, 'i titiro ahau i te tekoteko, aua.' She gave her head two shakes to emphasise that when she had looked at the tekoteko it didn't look like Paikea at all. Instead, the tekoteko looked like a human girl. 'He mokopuna na Paikea, pea?' she asked modestly. *'He mauri no Paikea?'* Her song inflected the questions with graceful ornamentation.

The other kuia nodded to each other. She was clever all right, the kai karanga. They were dumb by comparison. By asking questions she was enabling their leader to come to the decision she had already reached. No wonder she was the queen and they were the ladies in waiting.

The koroua waved the toa taiaha away; he was getting irritated with that toa taiaha and his fancy drill.

'He mauri?' the koroua repeated to himself. And through the mists of time he saw his master, Paikea, flinging wooden spears into the sky. Some in mid flight became birds. And others on reaching the sea turned into eels. And he, Paikea himself, was a mauri populating the land and sea so that it was no longer barren.

The koroua began to assess the weight of the tekoteko and, pae kare, it was light all right, and the legs were shorter than he remembered and —

'Ae,' the kai karanga crooned, agreeing with the decision he hadn't yet made, *'te mauri o Paikea tenei tekoteko.'* She let the words sink in, because she knew that it always took the males longer than the females to understand. She wanted to make sure that the koroua really understood that the tekoteko was a mauri and, if it was not returned to the surface and taken back to the land, then it would not fulfil its tasks. *'He mauri o Paikea,'* she said, *'mo te puawaitanga o te ure o Paikea.'* In her voice was ageless music.

The koroua swayed in the silken tides of the stirring sea. Though tired, he sensed the truth in his consort's words. For he remembered that Paikea had hesitated before throwing the last of his wooden spears and, when he did this, he had said, 'Let this one be planted in the years to come when the people are troubled and the mauri is most needed.' And the mauri, soaring through the sky came to rest in the earth where the afterbirth of a female child would be placed.

And as he remembered, the koroua began to lose his nostalgia for the past and to put his thoughts to the present and the future. Surely, in the tidal waves of Fate, there must have been a reason for his living so long. It could not have been coincidence that he should return to Whangara and be ridden by a descendant of his beloved golden master. Perhaps his fate and that of the tekoteko were inextricably intertwined? Ah yes, for nothing would have been left to chance.

The ope as they waited for the koroua's judgment began to add the colour of their opinion. The kuia chattered that they knew all along the kai karanga was right, and the toa taiaha, seeing the way things were going, added his two cents' worth also.

The koroua gave a swift gesture.

'Me piki mai tatou ki te rangi,' he commanded, readying himself for a quick ascent to the surface. 'Mauria mai ta tatou tekoteko ki te Ao o Tane. Kei te tautoko?'

The ope sang a waiata of agreement to their ancient leader's decision that they should carry the tekoteko back to the world of Man.

'Ae. Ae. Ae,' they chorused in a song of benign and burnished tenderness. 'Ae. Ae. Ae.'

Slowly, the phalanx of whales began their graceful procession to the surface of the sea, broadcasting their orchestral affirmation to the universe.

Hui e, haumi e, taiki e.

Twenty

After Kahu's departure, Nanny Flowers collapsed. She was taken to the hospital where, five days later, her eyelids flickered open. She saw Koro Apirana sitting next to the bed. Me and the boys were also there.

Nanny Flowers shook herself awake. The nurse and Koro Apirana helped her to sit up. Once she had gotten comfortable she closed her eyes a second time. Then she peeked out of one eye and sighed.

'Aue,' she said sarcastically. 'If you lot are still here that must mean I haven't gone to Heaven.'

But we didn't mind her sarcasm because we were used to her being an old grump. Koro Apirana looked at her lovingly.

'You have to lose some weight, Putiputi,' he said to her. 'Your ticker is too weak. I don't know what I would have done if the both of you —'

Nanny Flowers suddenly remembered. 'Aue, taku moko-puna, taku Kahu.'

Koro Apirana quietened her quickly. 'No, no, e kui,' he said. 'She's all right. She's all right.' He told Nanny Flowers what had happened.

Three days after the sacred whale and its accompanying herd had gone, and after Kahu had been given up for dead, she had been found unconscious, floating in a nest of dark lustrous kelp in the middle of the ocean. How she got there nobody knew, but when she was found the dolphins that were guarding her sped away with happy somersaults and leaps into the air.

Kahu had been rushed to the hospital. Her breathing had stopped, started, stopped and then started again. She was now off the respirator but she was still in a coma. The doctors did not know whether she would regain consciousness.

'Where is she? Where's my Kahu?' Nanny Flowers cried.

'She's here with you,' Koro Apirana said. 'Right here in this same hospital. Me and the iwi have been looking over you both, waiting for you to come back to us. You two have been mates for each other, just like in the kumara patch.'

Koro Apirana gestured to the other bed in the room. The boys separated and, through the gap, Nanny Flowers saw a little girl in pigtails, her face waxed and still.

The tears streamed down Nanny Flowers' cheeks.

'Push my bed over to her bed,' Nanny said. 'I'm too far away from her. I want to hold her and talk to her.'

The boys huffed and puffed with pretended exertion.

'Now all of you Big Ears can wait outside the door,' Nanny said. 'Just leave me and your Koro here alone with our mokopuna.'

She was like a little doll. Her eyes were closed and her eyelashes looked very long against her pallid skin. White

142

ribbons had been used to tie her plaits. There was no colour in her cheeks, and she seemed not to be breathing at all.

The bedcovers had been pulled right up to Kahu's chin, but her arms were on top of the covers. She was wearing warm flannel pyjamas, and the pyjama top was buttoned up to her neck.

The minutes passed. Koro Apirana and Nanny Flowers looked at each other, and their hearts ached.

'You know, dear,' Koro Apirana said, 'I blame myself for this. It's all my fault.'

'Yeah, it sure is,' Nanny Flowers wept.

'I should have known she was the one,' Koro Apirana said. 'Ever since that time when she was a baby and bit my toe.'

'Boy, if only she had real teeth,' Nanny Flowers agreed.

'And all those times I packed her away from the meeting house, I should have known.'

'You were deaf, dumb, blind *and* stubborn.'

The window to the room was half open. The sunlight shone through the billowing curtains. Nanny Flowers noticed that the door was slowly inching open and that the nosey parkers were looking in. Talk about no privacy, with them out there with their eyes all red and the hupe coming out.

'You never even helped with Kahu's pito,' Nanny Flowers sobbed.

'You're right, dear, I've been no good.'

'Always telling Kahu she's no use because she's a girl. Always growling her. Growl, growl, growl.'

'And I never knew,' Koro Apirana said, 'until you showed me the stone.'

'I should have cracked you over the head with it, you old paka.'

Dappled shadows chased each other across the white walls. On the window-sill were vases of flowers in glorious profusion.

Koro Apirana suddenly got up from his chair. His face was filled with the understanding of how rotten he had been.

'You should divorce me,' he said to Nanny Flowers. 'You should go and marry old Waari over the hill.'

'Yeah, I should too,' Nanny Flowers said. 'He knows how to treat a woman. He wouldn't trample on my Muriwai blood or my kuia as much as you have.'

'You're right, dear, you're right.'

'I'm always right, you old paka, and —'

Suddenly Kahu gave a long sigh. Her eyebrows began to knit as if she was thinking of something.

'You two are always arguing,' she breathed.

The whales were rising from the sea. Their skins were lucent and their profiles were gilded with the moon's splendour. Rising, rising.

'Kei te ora ia?' the koroua asked. He was concerned that the tekoteko was okay, still breathing.

'Kei te ora,' the kai karanga nodded. She had been singing gently to the whale rider, telling her not to be afraid.

'Kei te pai,' the koroua said. 'Kaatahi, kei te ora tonu tana iwi, tana kuia, me te koroua, me na tamariki.'

And the whale herd sang their gladness that the iwi would also live, because they knew that the girl would need to be carefully taught before she could claim the place for her people in the world.

The whales breached the surface and the thunderous spray was like silver fountains in the moonlight.

Twenty-one

Nanny Flowers gave an anguished sob and reached out to hold Kahu tightly. Koro Apirana tottered to the bedside and looked down at the sleeping girl. He began to say a prayer, a karakia, and he asked the Gods to forgive him. He saw Kahu stir.

Oh *yes*, mokopuna. Rise up from the depths of your long moe. Return to the people and take your rightful place among them.

Kahu drew another breath. She opened her eyes. 'Is it time to wake up now?' she asked.

Nanny Flowers began to blubber. Koro Apirana's heart skipped a beat. 'Ae. Kua tae mai te wa.'

'They told me not to wake until you were both here,' Kahu said gravely.

'Ko wai ratou?' Koro Apirana asked.

'The whales,' she said. Then she smiled, 'You two sounded just like the kai karanga and the koroua arguing.'

Nanny Flowers looked up at Koro Apirana. 'We don't argue,' she said. '*He* argues and *I* win.'

'Your Muriwai blood,' Koro Apirana said. 'Always too strong for me.'

Kahu giggled. She paused. Then her eyes brimmed with tears. In a small voice she said, 'I fell off.'

'He aha?'

'I fell off the whale. If I was a boy I would have held on tight. I'm sorry, Paka, I'm not a boy.'

The old man cradled Kahu in his arms, partly because of emotion and partly because he didn't want those big ears out there to hear their big chief crying.

'You're the best mokopuna in the whole wide world,' he said. 'Boy or girl, it doesn't matter.'

'Really, Paka?' Kahu gasped. She hugged him tightly and pressed her face against him. 'Oh thank you, Paka. You're the best koro in the whole wide world.'

'I love you,' Koro Apirana said.

'Me too,' Nanny Flowers added.

'And don't forget about us,' said the rest of the iwi as they crowded into the room.

Suddenly, in the joyous melee, Kahu raised a finger to her lips: *Sssshh.*

The koroua breached the surface, leaping high into the moonlit sky. The sacred sign, the moko, was agleam like liquid silver. The koroua flexed his muscles, releasing Kahu, and she felt herself tumbling along his back, tumbling, tumbling, tumbling. All around her the whales were leaping, and the air was filled with diamond spray.

'Can't you hear them?' Kahu asked. *Interlock.*

She fell into the sea. The thunder of the whales departing was loud in her ears. She opened her eyes and looked downward. Through the foaming water she could see huge tail fins waving farewell, 'E hine, e noho ra.'

Then from the backwash of Time came the voice of the koroua, the ancient bull whale. 'Child, your iwi await you. Return to the Kingdom of Tane and fulfil your destiny.' And suddenly the sea was drenched again with a glorious echoing music from the dark shapes sounding.

Kahu looked at Koro Apirana, her eyes shining.

'Oh *Paka*, can't you hear them? I've been listening to them for ages now. Oh *Paka*, and the whales are still singing,' she said.

Haumi e, hui e, *taiki e*.

New York
14 August 1986

Glossary

ae yes
ahau I, me
Ahuahu Mercury Island
ake own, self
ana *(used after verb)* implies
  movement
anei here it is
ano again
ao world
Aotearoa New Zealand
api oven
ara arise
aroha love
arohanui great love
atu away (from speaker)
atua first (archaic usage)
aua don't know
aue oh dear!
awhi (-tia) embrace
engari but, however
Eruera Edward
haere travel
haka war dance
hapuku groper *Polyprion
  oxygeneios*
hara sin
haramai come here
Hawaiki traditional homeland of
  the Maori people

hei to, as
Henare Henry
hika form of address to either
  sex
hine form of address to a girl
hoa partner
hoki indeed
hokowhitu war party
hongi press noses
huhu beetle grub *Prionoplus
  reticularis*
hui gathering
hupe mucus
huri turn
huware spittle
ia he, she, him, her
ihi power
ika fish
ingoa name
iwi tribe
kaha strength
kahawai fish *Arripis trutta*
kahu hawk, chief
kai food
kainga home
kakahu cloak
kanga pirau fermented corn
kani saw (*noun* and *verb*)
kaore to express surprise

karakia prayer
karanga call
kare not
katahi then
katoa all
kaukau swim
ke another
kei located at
kete basket
kia said, told
kingi king, imperial
kiwa dark
ko (used before proper names)
koe you (one person)
kohanga nursery
koia that's it
koka mother
koro sir
koroua old man
koruru cloudy
koutou you (pl.)
kowhaiwhai scroll painting on rafter
kua + *verb* *indicates completed action, as in* kua whati = is broken
kui ladybird
kuia old woman
kuini queen
kumu posterior
mai this direction
mako shark *Isurus glaucus*
mana prestige
manaaki hospitality
manawanui brave
manga barracouta *Thyrsites atun*
mango ururoa great white shark *Carcharodon carcharias*
Maori indigenous people
marae focal point of settlement

mate death
maua we two
mauri life principle
mauria take, to be taken
mihimihi introductory speech-making
mimi urinate
moana sea
moe doze
mokemoke lonely
moki trumpeter fish *Latridopsis* spp.
moko tattoo
mokopuna young generation
muriwai backwater
na belonging to, by
ne (*implies question*) isn't it?
nei this (*connected with person speaking*)
neke shift
nga the (pl.)
ngati people of
ngaua bitten
nikau palm *Rhopalostylis sapida*
noho remain
nui big
ope expedition
ora alive
pae circumference
pae kare by golly!
pai quality
paka bugger
Pakeha non-Maori
pakia touched
patupaiarehe fairy
paua shellfish *Haliotis* spp.
piki climb
pirau decompose
pito end, birth cord
piupiu flax skirt

pohutukawa native Christmas tree *Metrosideros excelsa*
porou eager
potiki youngest
pounamu greenstone
pouri grief
poutama steps
Poutu-te-rangi star Altair
puawaitanga blossoming
puku abdomen
putiputi flower
puwhakahara tree *Olea* spp.
ra sun
rahui embargo
rangatira noble
rangi sky
Rangitane tribe of Manawatu/ Wairarapa
ratou them (more than two)
Raukawa Cook Strait
Rawheoro site of traditional East Coast carving school
rawhiti east
Rawiri David
Rehua star Antares
reo speech
ringa arm, hand
Rotorua Bay of Plenty city
runga upwards
ta (*possessive*) the ... of
tahu plenty
taiaha long club
taku my (one item)
tama boy
tamahine girl
tamariki children
tamure fish *Pagrasomus auratus*
tana his, her, its
Tane Mahuta guardian of the forest

Tangaroa guardian of the sea
tangata person (either sex)
tangi mourn
taniwha water monster
tapere amusement
tapu sacred
tarawhai stingray
taringa ear
tatou us (including the one spoken to)
taunga landing place
tautoko to support
Tawhirimatea god of winds and storm
tawhiti distance
te the (*singular*)
tekoteko carved figure on a house
tena that (near you)
tenei this
tera saddle
tere speed
tiko defecate
tipua guardian spirit
tipuna ancestor
titiro look
toa warrior
tohora southern right whale *Baelena glacialis australis*
tohu emblem
tohunga specialist esp. artist, priest
toia drag
toka rock
tomo enter
tonu constant
toroa middle finger
tu stance
tuahine sister, female cousin (of a male)

Tuamotu   East Polynesian
   archipelago
tuatara   ancient reptile
   Sphenodon punctatus
turituri   shut up!
uia   be asked questions
umu   earth oven
ure   courage
wa   duration
waenganui   in the middle
waha   mouth
wahine   woman
wai   water
waiapu   fine flax kilt
waiata   song poem
Wainui   mother of heavenly
   bodies
waka   canoe
wananga   seminar

warehou   fish *Seriolelle brama*
wehenga   division
weka   wood-hen *Gallirallus
   australis*
whaiaipo   sweetheart
whakaeroero   putrefy
whakahuatia   be recited
whakapapa   genealogy
whakarongo   listen
whakatane   mannish (in a woman)
whanau   extended family
whare kai   dining room
whare   house
Whatonga   East Coast ancestor
wheke   octopus *O. maorum*
whenua   ground
whironui   ancestor
whiti (-a)   spring up
whiu (-a)   put in place